Springer Proceedings in Mathematics & Statistics

Volume 405

This book series features volumes composed of selected contributions from workshops and conferences in all areas of current research in mathematics and statistics, including data science, operations research and optimization. In addition to an overall evaluation of the interest, scientific quality, and timeliness of each proposal at the hands of the publisher, individual contributions are all refereed to the high quality standards of leading journals in the field. Thus, this series provides the research community with well-edited, authoritative reports on developments in the most exciting areas of mathematical and statistical research today.

Raffaele Argiento · Federico Camerlenghi ·
Sally Paganin
Editors

New Frontiers in Bayesian Statistics

BAYSM 2021, Online, September 1–3

Springer

Editors
Raffaele Argiento
Dipartimento di Scienze Economiche
Università degli Studi di Bergamo
Bergamo, Italy

Federico Camerlenghi
Department of Economics, Management and Statistics
University of Milano-Bicocca
Milan, Italy

Sally Paganin
Department of Biostatistics
Harvard T. H. Chan School of Public Health
Boston, MA, USA

ISSN 2194-1009 ISSN 2194-1017 (electronic)
Springer Proceedings in Mathematics & Statistics
ISBN 978-3-031-16426-2 ISBN 978-3-031-16427-9 (eBook)
https://doi.org/10.1007/978-3-031-16427-9

Mathematics Subject Classification: 62C10, 62F15, 62-06, 62Fxx, 62Gxx, 62Mxx, 62Pxx, 62Hxx

This Springer imprint is published by the registered company Springer Nature Switzerland AG
The registered company address is: Gewerbestrasse 11, 6330 Cham, Switzerland

Organization

Chair

Raffaele Argiento, Università degli Studi di Bergamo, Italy

Scientific Committee

Raffaele Argiento, Università degli Studi di Bergamo, Italy
Federico Camerlenghi, University of Milano-Bicocca, Italy
Alejandra Avalos-Pacheco, Harvard Medical School, USA
Roberta de Vito, Brown University, USA
Yanxun Xu, Johns Hopkins University, USA

Organizing Committee

Sally Paganin, Harvard University, USA
Willem van den Boom, National University of Singapore, Singapore
Jianjun Zhou, Yunnan University, China
Weixuan Zhu, Xiamen University, China
Tommaso Rigon, University of Milano-Bicocca, Italy
Daeyoung Lim, University of Connecticut, USA

Preface

This volume collects a selection of peer-reviewed contributions presented at the fifth Bayesian Young Statistician Meeting (BaYSM 2021), showcasing current and new advances in the frontiers of Bayesian statistics. The conference was originally planned to take place during 2020 at the Yunnan University, as a satellite to the ISBA 2020 World Meeting in Kunming, China. Due to the COVID-19 pandemic, the conference was first postponed and then turned into a virtual event held completely online from 1 to 3 September 2021. The event was patronized and graciously supported by the Yunnan University, the International Society of Bayesian Analysis (ISBA) and the junior section of ISBA (j-ISBA).

BaYSM is a conference designed as a platform for early-career researchers, including M.Sc. and Ph.D. students as well as post-doctoral researchers, to connect with the Bayesian scientific community at the beginning of their career. BaYSM aims at stimulating collaborations, encouraging discussion, and establishing networks between early-stage researcher as well as with senior professors.

The scientific program of BaYSM 2021 included six keynote sessions delivered by established senior researchers, in addition to contributed sessions and poster sessions featuring junior researchers. The keynote sessions offered brilliant and stimulating talks by Maria De Iorio (Yale-NUS College and University College London), David Dunson (Duke University), Long Nguyen (University of Michigan), Amy Shi (SAS), Jessica Utts (University of California, Irvine), and Francesca Dominici (Harvard University).

Contributions from early-career researchers highlighted different aspects of Bayesian statistics. Applications included personalized medicine to epidemiology and migrations pathways. Computation ranged from MCMC methods to ABC, while among methodological contributions, we mention graphical models and Bayesian nonparametrics. A senior discussant was present at each contributed session, providing helpful suggestions, encouragement, and directions for current and future research. One of the major challenges of this edition was the organization of an online environment able to facilitate virtual networking among participants. This goal was achieved setting up a virtual-town where participants could freely roam and chat with each other, altogether with the planning of a community event involving a Bayesian themed Quiz. Finally, through the support of our sponsors, outstanding contributions of junior researchers were acknowledged by four prizes recognizing the Best Talk and Best Poster in "Theory and Methods" and in "Applications and Computation", as well as two honourable mentions for Best Talk.

We acknowledge all BaYSM 2021 attendees, whose active (online) participation and contributions made the conference an amazing scientific event and an enjoyable experience. We thank the speakers, both junior and senior; and in particular, we are grateful to the discussants—Li Ma, Pierre Jacob, David Rossell, Rosangela H. Loschi, Christian Robert, Peter Müller—for their valuable work. Moreover, a special thanks goes to Michele Guindani for hosting the Bayesian themed Quiz in spite of the time zone difficulties. Finally, we express our sincere gratitude to the referees, who thoroughly reviewed the contributions in this volume and provided helpful comments for the early-career researchers.

Despite the challenges of the pandemic, organizing BaYSM 2021 has been an exciting and rewarding experience for all the committee's members. BaYSM has now become the official meeting of j-ISBA; we hope that the j-ISBA section will become more and more a point of reference for early-career researchers, and that the BaYSM conference will continue with the same success of the previous editions, providing inspiration for new generations of Bayesian statisticians. We look forward for the next in-person meeting, which will be held next month in Montréal, Canada, as a satellite event of the ISBA 2022 World Meeting.

Bergamo, Italy
Milan, Italy
Boston, USA
May 2022

Raffaele Argiento
Federico Camerlenghi
Sally Paganin

Contents

About the Editors

Raffaele Argiento is a full professor of Statistics at the Department of Economics of the University of Bergamo, Italy. He is a member of Economics, Statistics, and Data Science Ph.D. board at the University of Milano-Bicocca, and he is affiliated to the "de Castro" Statistics initiative of the Collegio Carlo Alberto, Turin. His research focuses on Bayesian finite and infinite mixture models with a particular focus on the associated computational strategies as well as the related model-based clustering.

Federico Camerlenghi is an assistant professor of Statistics at the Department of Economics, Management, and Statistics and a board member of the Ph.D. in Economics, Statistics, and Data Science at the University of Milano-Bicocca, Italy. His research mainly focuses on the construction and investigation of Bayesian nonparameteric models to handle exchangeable and partially exchangeable data. He received the Savage award in Theory and Methods in 2017, and he has been the chair of the Junior Section of the International Society for Bayesian Analysis (j-ISBA) in 2019.

Sally Paganin is a research fellow in the Department of Biostatistics at Harvard T. H. Chan School of Public Health and treasurer of the Junior Section of the International Society for Bayesian Analysis (j-ISBA). Previously, she was a postdoctoral researcher at UC Berkeley and a core team member of the NIMBLE software project. Her research focuses on Bayesian methods and statistical models for complex data, along with the development of statistical software and algorithms.

Approximate Bayesian Algorithm for Tensor Robust Principal Component Analysis

Andrej Srakar

Abstract Recently proposed Tensor Robust Principal Component Analysis (TRPCA) (Lu et al. in Tensor robust principal component analysis: exact recovery of corrupted low-rank tensors via convex optimization, 2019 [14]) aims to exactly recover the low-rank and sparse components from their sum, extending the Low-Rank Tensor Completion model of Mu et al. (Lower bounds and improved relaxations for tensor recovery, 2013 [17]). We construct a Bayesian approximate inference algorithm for TRPCA, based on regression adjustment methods suggested in the literature to correct for high-dimensional nature of the problem (Blum in J Am Stat Assoc 105(491), 2010 [3]; Blum and François in Stat Comput 20(1):63–73, 2010 [4]). Our results are compared to previous studies using variational Bayes inference for tensor completion (Hawkins and Zhang in Conference: IEEE international conference on data mining, 2018 [11]). In a short application, we study spatiotemporal traffic data imputation using nine-week spatiotemporal traffic speed data set of Guangzhou, China.

Keywords Tensor robust PCA · Low-rank · Tensor completion · Approximate bayesian computation · Regression adjustment · Variational bayes

AMS Subject Classification 62F15

1 Introduction: Tensor Robust Principal Component Analysis and Its Extensions

Classical Principal Component Analysis (PCA) is the most widely used statistical tool for data analysis and dimensionality reduction. It is computationally efficient and powerful for the data which are mildly corrupted by small noises. However, a major issue of PCA is that it is brittle to grossly corrupted observations or presence of outliers, which are ubiquitous in real world data. To date, a number of robust ver-

A. Srakar (✉)
Ljubljana and School of Economics and Business, Institute for Economic Research (IER), University of Ljubljana, Ljubljana, Slovenia
e-mail: andrej.srakar@ier.si

R. Argiento et al. (eds.), *New Frontiers in Bayesian Statistics*, Springer Proceedings in Mathematics & Statistics 405, https://doi.org/10.1007/978-3-031-16427-9_1

sions of PCA were proposed. But many of them suffer from the high computational cost. The recently proposed Robust PCA [7] is the first polynomial-time algorithm with strong performance guarantees. Suppose we are given a data matrix $X \in \mathbb{R}^{n_1 \times n_2}$ which can be decomposed as $X = L_0 + E_0$ where L_0 is low-rank and E_0 is sparse. It is shown in Candès et al. [7] that if the singular vectors of L_0 satisfy some incoherent conditions, L_0 is low-rank and E_0 is sufficiently sparse, then L_0 and E_0 can be recovered with high probability by solving the following convex optimization problem:

$$\min_{L,E} \|L\|_* + \lambda \|E\|_1, \quad s.t.\ X = L + E \tag{1}$$

where $\|L\|_*$ denotes the nuclear norm (sum of the singular values of L), $\|E\|_1$ denotes the $\ell 1$-norm (sum of the absolute values of all the entries in E) and

$$\lambda = 1/\sqrt{\max(n_1, n_2)} \tag{2}$$

To use RPCA, one has to first restructure/transform the multi-way data into a matrix. Such a preprocessing usually leads to the information loss and would cause performance degradation. To alleviate this issue, a common approach is to manipulate the tensor data by taking the advantage of its multi-dimensional structure. In this work, we study the Tensor Robust Principal Component (TRPCA) which aims to exactly recover a low-rank tensor corrupted by sparse errors.

Tensors are mathematical objects that can be used to describe physical properties, just like scalars and vectors. They are a generalisation of scalars and vectors; a scalar is a zero rank tensor, and a vector is a first rank tensor. The rank (or order) of a tensor is defined by the number of directions (i.e. dimensionality of the array) required to describe it.

The tensor multi rank of $\mathcal{A} \in \mathbb{R}^{n_1 \times n_2 \times n_3}$ is a vector $r \in \mathbb{R}^{n_3}$ with its i-th entry as the rank of the i-th frontal slice of $\overline{\mathcal{A}}$, i.e., $r_i = rank(\overline{\mathcal{A}}^{(i)})$. The tensor tubal rank, denoted as $\text{rank}_t(\mathcal{A})$, is defined as the number of nonzero singular tubes of $\mathcal{S}$, where $\mathcal{S}$ is from the t-SVD of $\mathcal{A} = \mathcal{U} * \mathcal{S} * \mathcal{V}^*$. That is

$$\text{rank}_t(\mathcal{A}) = \#\{i : \mathcal{S}(i, i, :) \neq 0\} = \max_i r_i \tag{3}$$

The tensor nuclear norm of a tensor $\mathcal{A} \in \mathbb{R}^{n_1 \times n_2 \times n_3}$, denoted as $\|\mathcal{A}\|_*$, is defined as the average of the nuclear norm of all the frontal slices of $\overline{\mathcal{A}}$, i.e., $\|\mathcal{A}\|_* = \frac{1}{n_3} \sum_{i=1}^{n_3} \left\| \overline{\mathcal{A}}^{(i)} \right\|_*$.

Tensor Robust PCA (TRPCA) [14] aims to exactly recover a low-rank tensor corrupted by sparse errors. It aims to recover the low tubal rank component $\mathcal{L}_0$ and sparse component $\mathcal{E}_0$ from $\mathcal{X} = \mathcal{L}_0 + \mathcal{E}_0 \in \mathbb{R}^{n_1 \times n_2 \times n_3}$ by convex optimization

$$\min_{\mathcal{L},\mathcal{E}} \|\mathcal{L}\|_* + \lambda \|\mathcal{E}\|_1, \ s.t. \mathcal{X} = \mathcal{L} + \mathcal{E} \tag{4}$$

We firstly define few necessary concepts. An *orthogonal tensor* is a tensor $Q \in \mathbb{R}^{n \times n \times n_3}$ if it satisfies:

$$Q^* * Q = Q * Q^* = I \tag{5}$$

where I is the identity tensor.
f-diagonal tensor is a tensor if each of its frontal slices is a diagonal matrix.
The Tensor Singular Value Decomposition (T-SVD) for third order tensors was proposed by Kilmer and Martin [13] and has been applied successfully in many fields, such as computed tomography, facial recognition, and video completion. Kilmer and Martin presented the concept of a tensor-tensor product with suitable algebraic structure such that classical matrix-like factorizations are possible. In particular, they gave the definition of the Tensor SVD (T-SVD) over this new product, and showed that truncating that expansion does give a compressed result that is optimal in the Frobenius norm.

Theorem 1 (Tensor Singular Value Decomposition (T-SVD) [13, 14]) *Let $\mathcal{A} \in \mathbb{R}^{n_1 \times n_2 \times n_3}$. Then it can be factored as:*

$$\mathcal{A} = \mathcal{U} * \mathcal{S} * \mathcal{V}^* \tag{6}$$

where $\mathcal{U} \in \mathbb{R}^{n_1 \times n_1 \times n_3}$, $\mathcal{V} \in \mathbb{R}^{n_2 \times n_2 \times n_3}$ are orthogonal, and $\mathcal{S} \in \mathbb{R}^{n_1 \times n_2 \times n_3}$ is an f-diagonal tensor.

Alternative tensor factorization is CANDECOMP/PARAFAC (CP) and expresses a N-way tensor $\mathcal{A}$ as the sum of multiple rank-1 tensors:

$$\mathcal{A} = \sum_{r=1}^{R} s_r a_r^{(1)} \circ \ldots \circ a_r^{(N)}, \quad with\ a_r^{(k)} \in \mathbb{R}^{I_k} \tag{7}$$

Our Bayesian approach is based on likelihood representation of the problem in (4) following variational Bayes perspective of Hawkins and Zhang [11]. Variational perspectives have been earlier adopted as solutions to intractable likelihood problems in matrix and tensor completion [1, 21, 22]. In general, likelihood free perspective is applied to matrix and tensor completion problems as computational complexity is significantly higher for high-dimensional data than that of other methods, and convergence is generally hard to assess [1, 5]. In order to address problems of high-dimensionality in approximate Bayesian inference regression adjustment is often recommended [2–5, 18] and we use it also in our analysis.

We assume that each tensor slice can be fit by $\mathcal{X}_k = \widetilde{\mathcal{X}}_k + \mathcal{S}_k + \mathcal{E}_k$, where $\widetilde{\mathcal{X}}_k$ is low-rank, $\mathcal{S}_k$ contains sparse outliers and $\mathcal{E}_k$ denotes dense noise with small magnitudes. We will denote with $\mathcal{Y}_{\Omega,k}$ the observation of current slice and by $\mathcal{S}_{\Omega,k}$ its outliers. For the likelihood function representation let τ specify the noise precision, $\widehat{a}_{i_n}^{(n)}$ the i_n-th row of $A^{(n)}$, λ controls the rank of factorization and $\{\gamma_{i_1,\ldots,i_N}\}$ controls the sparsity of $\mathcal{S}_\Omega$.

We define the likelihood function and used priors for the transformed problem in (4) using Gaussian and Gamma priors as:

$$\left(\mathcal{Y}_\Omega \left| \left\{A^{(n)}\right\}_{n=1}^{N+1}, \mathcal{S}_\Omega, \tau\right.\right) = \prod_{(i_1,\ldots,i_n)\in\Omega} \mathcal{N}\left(\mathcal{Y}_{i_1\ldots i_N} \left| \left\langle \widehat{a}_{i_1}^{(1)}, \ldots, \widehat{a}_{i_N}^{(N)} \right\rangle + \mathcal{S}_{i_1\ldots i_N}, \tau^{-1}\right.\right) \tag{8}$$

$$(\Theta\,|\mathcal{Y}_\Omega) = \frac{p\left(\mathcal{Y}_\Omega \left| \left\{A^{(n)}\right\}_{n=1}^{N+1}, \mathcal{S}_\Omega, \tau\right.\right) \left\{\prod_{n=1}^{(N+1)} p\left(A^{(n)}|\lambda\right)\right\} p\left(\lambda\right) p\left(\mathcal{S}_\Omega|\gamma\right) p\left(\gamma\right) p(\tau)}{p(\mathcal{Y}_\Omega)} \tag{9}$$

$$p\left(A^{(n)}\,|\lambda\right) = \prod_{i_n=1}^{I_n} \mathcal{N}\left(\widehat{a}_{i_n}^{(n)}\,|0, \Lambda^{-1}\right), \quad \forall n \in [1, N+1] \tag{10}$$

$$p\left(\mathcal{S}_\Omega \mid \gamma\right) = \prod_{(i_1,\ldots,i_N)\in\Omega} \mathcal{N}\left(\mathcal{S}_{i_1\ldots i_N}\,|0, \gamma_{i_1\ldots i_N}^{-1}\right) \tag{11}$$

$$p\left(\tau\right) = Ga\left(\tau\,|a_0^\tau, b_0^\tau\right), \quad p\left(\lambda\right) = \prod_{r=1}^{R} Ga\left(\lambda_r|c_0, d_0\right) \tag{12}$$

$$p(\gamma) = \prod_{(i_1,\ldots,i_N)\in\Omega} Ga\left(\gamma_{i_1\ldots i_N}\,|a_0^\gamma, b_0^\gamma\right) \tag{13}$$

2 Scheme of the Approximate Bayesian Algorithm

Modern statistical applications increasingly require the fitting of complex statistical models. Often these models are "intractable" in the sense that it is impossible to evaluate the likelihood function. This prohibits standard implementation of likelihood-based methods, such as maximum likelihood estimation or a Bayesian analysis. To overcome this problem there has been substantial interest in "likelihood-free" or simulation-based methods. Examples of such likelihood-free methods include simulated methods of moments [10], indirect inference (Gourièroux and Ronchetti 1993) [12], synthetic likelihood [9] and approximate Bayesian computation [19]. Of these, approximate Bayesian computation (ABC) methods are arguably the most common methods for performing Bayesian inference [15, 19]. For a number of years, ABC methods have been popular in population genetics (e.g. Cornuet et al. [8]) and systems biology (e.g. Toni et al. [20]); more recently they have seen increased use in other application areas, such as econometrics [6] and epidemiology [9].

In our ABC algorithm for TRPCA we amend the variational Bayes perspective of Hawkins and Zhang [11] who use it on a temporally defined problem. We use regression adjustment based ABC using as a summary statistic array of tensor first and second moment defined as k-statistics [16] and tensor tubal rank as defined above.

Scheme of the algorithm:

Step 1. Simulate $\theta^{(i)}, i = 1, \ldots, n$ according to the prior structure defined above.
Step 2. Simulate $s^{(i)} = \text{array}(\mathcal{A})^{(i)}$ using the generative model $p(s^{(i)}|\theta^{(i)})$.
Step 3. Associate with each pair $(\theta^{(i)}, s^{(i)})$ a weight $w^{(i)} \propto K_h(\|s^{(i)} - s_{\text{obs}}\|)$, where K_h is a kernel function and $\|\ \|$ the multidimensional Euclidean distance.
Step 4. Fit a regression model where the response is θ and the predictive variables are the summary statistics s. Use a regression model to adjust the $\theta^{(i)}$ in order to produce a weighted sample of adjusted values. We use heteroskedastic adjustment, following Blum (2017), as follows:

$$\theta_{c'}^{(i)} = \widehat{m}(s_{\text{obs}}) + \frac{\widehat{\sigma}(s_{\text{obs}})}{\widehat{\sigma}(s^{(i)})}(\theta^{(i)} - \widehat{m}(s^{(i)})) \tag{14}$$

where $\widehat{m}$ and $\widehat{\sigma}$ are the standard estimators of the conditional mean and of the conditional standard deviation.

3 Numerical Experiments and Application

With the development of intelligent transportation systems, large quantities of urban traffic data are collected on a continuous basis from various sources. These data sets capture the underlying states and dynamics of transportation networks and the whole system. In general, traffic data register full spatial and temporal features, together with some other site-specific attributes. Usually, we can organize the spatiotemporal traffic data into a multi-dimensional structure. Combined with information from other links in a city, the overall spatiotemporal data can be structured as a multi-dimensional array, which is often referred to as a tensor. A common drawback that undermines the use of such spatiotemporal data is the "missingness" problem, which may result from various factors such as hardware/software failure, network communication problems, and zero/limited reports from floating/crowdsourcing systems.

To demonstrate the performance of this model, in this section we conduct numerical experiments based on a large-scale traffic speed data set collected in Guangzhou, China. The data set is generated by a widely-used navigation app on smart phones. The data set contains travel speed observations from 214 road segments in two months (61 days from August 1, 2016 to September 30, 2016) at 10-min interval (144 time intervals in a day). The speed data can be organized as a third-order tensor (road segment × day × time interval). Among the 1.88 million elements, about 1.29% are not observed or provided in the raw data.

In Tables 1 and 2 we compare performance of different models applied to several scenarios. We compare: Bayesian Gaussian CANDECOMP/PARAFAC (BGCP) tensor decomposition model, high accuracy low-rank tensor completion (HaLRTC) (Liu et al. 2013), which is used in Ran et al. (2016), SVD-combined tensor decomposition (STD) (Chen et al. 2018), DA (daily average) fills the missing value with an average of observed data (over different days) for the same road segment and the same time window (Li et al. 2013). kNN is another baseline method where the neighbors refer to road segments. Finally, TRPCA-VAR and TRPCA-ABC refer to tensor robust PCA specification in variational Bayes and approximate Bayesian computation algorithm form. The mean absolute percentage error (MAPE) and root mean square error (RMSE) are used to evaluate model performance. Our first experiment examines the performance of different models and different representations in the random missing scenario. In the second experiment, we present a more realistic temporally correlated missing scenario. From the original data set we create five novel datasets with different missing rates ranging from 10 to 50%. We use two data representations: matrix representation (A) and third-order tensor representation (B).

As can be seen from the tables (the best models are marked in bold), for the random missing scenario, frequently the Variational Bayes specification performs best. On the other hand, our ABC approach performs very well in the second, temporally correlated missing scenario.

4 Conclusion

Our article provides an initial step in the development of ABC algorithms for tensor completion and tensor principal component analysis. We upgrade the tensor robust PCA approach of Lu and coauthors using approximate Bayesian perspective which provides ground for further research in the area of Bayesian approaches in matrix and tensor completion. Also, our article provides additional information on approximate Bayesian approaches to high-dimensional problems in statistics.

Few possible extensions of our work and pathways for future work seem apparent:

- Other possibilities of the ABC algorithms (such as SMC, HMC, other regression and marginal adjustment approaches) integrated nested Laplace approximation, including additional upgrades of the variational approach of Hawkins and Zhang should lead to more evidence on methodological possibilities to approach matrix and tensor completion from a Bayesian computational perspective.
- Different loss and divergence measures (for example Bregman type divergence measures) could be tested and asymptotics of the approach developed.
- Extension to different type of tensor measures and different specifications of the tensor robust PCA (the specification we use is only one of the possible ones) as well as extensions to any type and size of a tensor.

Table 1 The imputation performance of BGCP, HaLRTC, STD, DA (daily average), kNN, TRPCA-VAR and TRPCA-ABC for two data representations in the first scenario (best models are highlighted in bold)

		10%		20%		30%		40%	
		MAPE	RMSE	MAPE	RMSE	MAPE	RMSE	MAPE	RMSE
(A)	BGCP(50)	0.0937	3.9981	0.0952	4.0467	0.0962	4.0903	0.0976	4.1457
	BGCP(80)	0.0925	3.9483	0.0941	3.9958	0.0951	4.0449	0.0968	4.1091
	BGCP(110)	0.0920	3.9303	0.0937	3.9790	0.0948	4.0292	0.0965	4.0937
	HaLRTC	0.0957	3.9666	0.0976	4.0232	0.0991	4.0820	0.1009	4.1467
	DA	0.1213	5.1778	0.1218	5.1905	0.1217	5.1977	0.1217	5.1993
	kNN(10)	0.1303	5.1101	0.1314	5.1486	0.1322	5.1966	0.1333	5.2565
	TRPCA-VAR	0.1003	4.3452	0.0786	4.4013	0.1045	4.4291	0.0948	4.4778
	TRPCA-ABC	0.1083	4.3652	0.1026	4.4063	0.0825	4.4141	0.1008	4.4854
(B)	BGCP(50)	0.0862	3.7097	0.0867	3.7199	0.0867	3.7298	0.0867	3.7317
	BGCP(80)	0.0823	3.5614	0.0827	3.5660	0.0827	3.5775	0.0829	3.5851
	BGCP(110)	0.0795	3.4521	0.0798	3.4531	0.0799	3.4655	**0.0801**	**3.4756**
	HaLRTC	0.0777	3.1917	0.0815	3.3324	0.0850	3.4748	0.0887	3.6143
	STD	0.0888	3.7708	0.0911	3.8308	0.0936	3.9286	0.0963	4.0265
	TRPCA-VAR	**0.0639**	**3.5291**	**0.0804**	**3.5534**	**0.0666**	**3.6102**	0.0929	3.6656
	TRPCA-ABC	0.0939	3.5441	0.0964	3.5924	0.0856	3.6332	0.0979	3.6886

Table 2 The imputation performance of BGCP, HaLRTC, STD, DA, kNN, TRPCA-VAR and TRPCA-ABC for two data representations in the second scenario (best models are highlighted in bold)

		10%		20%		30%		40%	
		MAPE	RMSE	MAPE	RMSE	MAPE	RMSE	MAPE	RMSE
(A)	BGCP(15)	0.1011	4.2458	0.1013	4.2674	0.1020	4.3162	0.1031	4.3915
	BGCP(20)	0.1005	4.2307	0.1010	4.2755	0.1017	4.3229	0.1031	4.4124
	HaLRTC	0.1015	**4.1322**	0.1022	4.1716	0.1035	4.2372	0.1057	4.3232
	DA	0.1208	5.1128	0.1207	5.1353	0.1200	5.1408	0.1196	5.1434
	kNN(13)	0.1342	5.1714	0.1340	5.1983	0.1346	5.2591	0.1388	5.4405
	TRPCA-VAR	0.0906	4.5926	0.1158	4.5866	0.1204	4.6352	0.1291	4.7332
	TRPCA-ABC	0.1226	4.5736	0.0978	4.5866	0.1264	4.6362	0.1001	4.7352
(B)	BGCP(15)	0.0992	4.1760	0.0995	4.1949	0.0999	4.2425	0.1001	**4.2881**
	BGCP(20)	0.0980	4.1413	0.0984	**4.1477**	**0.0980**	**4.1857**	0.1006	4.4556
	HaLRTC	0.1033	4.1576	0.1046	4.2086	0.1062	4.2792	0.1088	4.3813
	STD	0.1019	4.1881	0.1054	4.3300	0.1068	4.4029	0.1115	4.5573
	TRPCA-VAR	0.0956	4.1598	0.1100	4.1903	0.1047	4.2626	0.0943	4.4116
	TRPCA-ABC	**0.0746**	4.1758	**0.0740**	4.2343	0.1167	4.2866	**0.0853**	4.4006

References

1. Babacan, S.D., Luessi, M., Molina, R., Katsaggelos, A.K.: Sparse Bayesian methods for low-rank matrix estimation. IEEE Trans. Signal Process **60**(8), 3964–3977 (2012)
2. Beaumont, M.A., Zhang, W., Balding, D.J.: Approximate Bayesian computation in population genetics. Genetics **162**, 2025–2035 (2002)
3. Blum, M.G.: Approximate Bayesian computation: a nonparametric perspective. J. Am. Statistical Assoc
4. Blum, M.G., François, O.: Non-linear regression models for approximate Bayesian computation. Stat. Comput. **20**(1), 63–73 (2010)
5. Blum, M.G.B., Nunes, M.A., Prangle, D., Sisson, S.A.: A comparative review of dimension reduction methods in approximate Bayesian computation. Stat. Sci. **28**, 189–208 (2013)
6. Calvet, L.E., Czellar, V.: Accurate methods for approximate Bayesian computation filtering. J. Financ. Econ. **13**(4), 798–838 (2015)
7. Candès, E.J., Li, X.D., Ma, Y., Wright, J.: Robust principal component analysis? J. ACM **58**(3) (2011)
8. Cornuet, J.-M., Santos, F., Beaumont, M.A., Robert, C.P., Marin, J.-M., Balding, D.J., Guillemaud, T., Estoup, A.: Inferring population history with DIY ABC: a user-friendly approach to approximate Bayesian computation. Bioinformatics **24**(23), 2713–2719 (2008)
9. Drovandi, C.C., Pettitt, A.N.: Estimation of parameters for macroparasite population evolution using approximate Bayesian computation. Biometrics **67**(1), 225–233 (2011)
10. Duffie, D., Singleton, K.J.: Simulated moments estimation of Markov models of asset prices. Econometrica **61**(4), 929–952 (1993)
11. Hawkins, C., Zhang, Z.: Variational Bayesian inference for robust streaming tensor factorization and completion. In: Conference: IEEE International Conference on Data Mining, Nov 2018. Available as arXiv:1809.01265v1 (2018)
12. Heggland, K., Frigessi, A.: Estimating functions in indirect inference. J. Roy. Stat. Soc. Ser. B **66**, 447–462 (2004)
13. Kilmer, M.E., Martin, C.D.: Factorization strategies for third-order tensors. Linear Algebra Appl. **435**(3), 641–658 (2011)
14. Lu, C., Feng, J., Chen, Y., Liu, W., Lin, Z., Yan, S.: Tensor robust principal component analysis: exact recovery of corrupted low-rank tensors via convex optimization. Available as arXiv:1708.04181v3 (2019)
15. Martin, G., Frazier, D., Robert, C.P.: Computing Bayes: Bayesian computation from 1763 to the 21st century. Available as arXiv:2004.06425 (2020)
16. McCullagh, P.: Tensor Methods in Statistics, 2nd edn. Dover Books on Mathematics (2018)
17. Mu, C., Huang, B., Wright, J., Goldfarb, D.: Square deal: lower bounds and improved relaxations for tensor recovery. Available as arXiv:1307.5870v2 (2013)
18. Nott, D.J., Ong, V.M.-H., Fan, Y., Sisson, S.A.: High-dimensional ABC. In: Handbook of Approximate Bayesian Computation, pp. 211–242 (2018)
19. Robert, C.P.: Approximate Bayesian computation: a survey on recent methods. In: Cools, R., Nuyens, D. (eds.) Monte Carlo and Quasi-Monte Carlo Methods (MCqMC), pp. 195–205. Springer, Berlin (2014)
20. Toni, T., Welch, D., Strelkowa, N., Ipsen, A., Stumpf, M.P.: Approximate Bayesian computation scheme for parameter inference and model selection in dynamical systems. J. Roy. Soc. Interface **6**(31), 187–202 (2009)
21. Zhao, Q., Zhang, L., Cichocki, A.: Bayesian CP factorization of incomplete tensors with automatic rank determination. IEEE Trans. Pattern Anal. Mach. Intell. **37**(9), 1751–1763 (2015)
22. Zhao, Q., Zhou, G., Zhang, L., Cichocki, A., Amari, S.-I.: Bayesian robust tensor factorization for incomplete multiway data. IEEE Trans. Neural Networks Learn. Syst. **27**(4), 736–748 (2016)

Bayesian Quantile Regression for Big Data Analysis

Yuanqi Chu, Xueping Hu, and Keming Yu

Abstract Quantile regression, which estimates various conditional quantiles of a response variable, including the median (0.5th quantile), is particularly useful when the conditional distribution is asymmetric or heterogeneous or fat-tailed or truncated. Bayesian methods for the inference of quantile regression have been receiving increasing attention from both theoretical and empirical viewpoints but facing the challenge of scaling up the existing methods when the data are too large to be processed by a single machine under many big data environments nowadays. In this paper, we explore Bayesian quantile regression (BQR) analysis via normal-inverse-gamma (NIG) distribution type of likelihood function, prior distribution and posterior distribution. We further develop the details of methods of BQR for massive data applications. The performance of proposed methods is evaluated via real data illustrations.

Keywords Quantile regression (QR) · Bayesian inference · Big data · Normal-inverse-gamma (*NIG*)

1 Introduction

Quantile regression (QR) estimates various conditional quantiles of a response or dependent random variable, including the median (0.5th quantile). Putting different quantile regressions together provides a more complete description of the underlying conditional distribution of the response than a simple mean regression. This is

Y. Chu · K. Yu (✉)
Department of Mathematics, Brunel University London, Middlesex UB8 3PH, UK
e-mail: keming.yu@brunel.ac.uk

Y. Chu
e-mail: yuanqi.chu@brunel.ac.uk

X. Hu · K. Yu
College of Mathematics and Physics, Anqing Normal University, Anqing 246133, People's Republic of China
e-mail: hxprob@163.com

R. Argiento et al. (eds.), *New Frontiers in Bayesian Statistics*, Springer Proceedings in Mathematics & Statistics 405, https://doi.org/10.1007/978-3-031-16427-9_2

particularly useful when the conditional distribution is asymmetric or heterogeneous or fat-tailed or truncated. Quantile regression has been widely used in statistics and numerous application areas ([3, 5, 11, 25] and among others). In the "big data" era for statistical science, the rich of data sources with many complicated data structures and the increase of extreme values and heterogeneity may see quantile regression methods more relevant than mean regression to dig deep into the data and grab information from it. In particular, with advanced power of computer, complicated quantile regression-based models could be developed under a Bayesian framework, and Bayesian quantile regression (BQR) has received increasing attention from both theoretical and empirical viewpoints with wide applications and variants (see [4, 10, 12, 17, 19, 23] and among others). So far, in the context of quantile regression, several methods have been developed for big data analysis ([6, 9, 22, 27] and among others), but little attention has been paid to such methodology under Bayesian inference paradigm.

In this paper, we propose a new approach of BQR for big data. This approach has its posterior distribution on the whole data as a joint posterior from M sub data split from the whole data. Section 2 introduces the likelihood function for BQR based on the location-scale mixture of normals for asymmetric Laplace distribution [15, 18]. Section 3 gives details of the normal-inverse-gamma (NIG) expressions of the prior and posterior distributions for BQR via informative g-prior [28]. Section 4 derives the posterior distribution on the whole data as a joint multiplication of the posterior obtained from M sub data split from the whole data via NIG summation operator, and provides big data based algorithms for BQR. Section 5 demonstrates the proposed approaches and algorithms via real data illustrations. Some concluding remarks are presented in Sect. 6.

2 Quantile Regression and Its Likelihood Function

Let $y_i, i = 1, \ldots, n$ be a continuous response variable and $\mathbf{x_i}$ a $k \times 1$ vector of predictors for the ith observation. The linear quantile regression model for the pth quantile can be denoted as $y_i = \mathbf{x_i}^T \boldsymbol{\beta} + \varepsilon_i$, where $\boldsymbol{\beta}$ is a $k \times 1$ vector of unknown parameters of interest, and ε_i is the error term whose distribution is assumed to have zero pth quantile. The estimation for $\boldsymbol{\beta}$ is solved by minimizing $\sum_{i=1}^n \rho_p(y_i - \mathbf{x_i}^T \boldsymbol{\beta})$, where $\rho_p(u) = u\{p - I(u < 0)\}$ is the check function and $I(\cdot)$ denotes the indicator function. According to [24, 26], such minimization is equivalent to maximizing a likelihood function that is based on the asymmetric Laplace distribution (ALD) at specific value of p. Assume that errors $\varepsilon_i, i = 1, \ldots, n$ are $ALD(0, \sigma, p)$, with the likelihood given by

$$f(\boldsymbol{\varepsilon}|\sigma) \propto \sigma^{-n} \exp\{-\sum_{i=1}^{n} \frac{|\varepsilon_i| + (2p-1)\varepsilon_i}{2\sigma}\},$$

where $\boldsymbol{\varepsilon} = (\varepsilon_1, \ldots, \varepsilon_n)^T$. Following [15, 18], we can represent ε_i as a location-scale mixture of normals as follows:

$$\varepsilon_i | v_i, \sigma \sim N((1-2p)v_i, 2\sigma v_i),\ v_i|\sigma \sim \mathrm{Exp}(\sigma^{-1}p(1-p)),$$

where $\mathrm{Exp}(\theta)$ denotes an exponential distribution with rate parameter θ. Denote $\mathbf{Y}$ as an $n \times 1$ response vector of y_i, $\mathbf{X}$ an $n \times k$ predictor matrix with ith row $\mathbf{x_i}^T$, we have

$$\mathbf{Y}|\boldsymbol{\beta}, \sigma, \mathbf{v}, \mathbf{X}, \boldsymbol{\Sigma} \sim N_n(\mathbf{X}\boldsymbol{\beta} + (1-2p)\mathbf{v}, 2\sigma\boldsymbol{\Sigma}),$$

where $\mathbf{v} = (v_1, \ldots, v_n)^T$ and $\boldsymbol{\Sigma}$ is the diagonal matrix of v_i. Given $\boldsymbol{\Sigma}$ and further let $\mathbf{Y}_p^* = \frac{1}{\sqrt{2}}(\mathbf{Y} - (1-2p)\mathbf{v})$, $\mathbf{X}^* = \frac{1}{\sqrt{2}}\mathbf{X}$ respectively, then $\mathbf{Y}_p^*$ follows a normal-type of conditional likelihood as

$$f(\mathbf{Y}_p^*|\boldsymbol{\beta}, \sigma, \mathbf{v}, \mathbf{X}^*, \boldsymbol{\Sigma}) \propto \sigma^{-n/2} \exp\{-\frac{1}{2\sigma}[\mathbf{Y}_p^* - \mathbf{X}^*\boldsymbol{\beta}]^T \boldsymbol{\Sigma}^{-1} [\mathbf{Y}_p^* - \mathbf{X}^*\boldsymbol{\beta}]\}. \quad (1)$$

3 *NIG* Prior and Posterior Distributions for Bayesian Quantile Regression

Mathematically, we introduce the definition of NIG [7] as follows.

Definition 1 Let $\boldsymbol{\beta}$ be a k-dimensional vector satisfying $-\infty < \boldsymbol{\beta} < \infty$ and $\delta > 0$ be the scalar parameter. The joint distribution of $(\boldsymbol{\beta}, \delta)$ follows the k-dimensional distribution $NIG_k(\boldsymbol{\mu}, \boldsymbol{\Lambda}, a, b)$ if

$$f(\boldsymbol{\beta}, \delta) = C\delta^{-(a+\frac{k}{2}+1)} \exp\{-\frac{1}{\delta}[b + \frac{1}{2}(\boldsymbol{\beta} - \boldsymbol{\mu})^T \boldsymbol{\Lambda}(\boldsymbol{\beta} - \boldsymbol{\mu})]\},$$

where C is a proportionality constant. That is, $f(\delta)$ follows the inverse-gamma (IG) distribution with shape parameter a and scale parameter b, and $f(\boldsymbol{\beta}|\delta)$ follows the multivariate normal distribution with $k \times 1$ mean vector $\boldsymbol{\mu}$ and $k \times k$ precision matrix $\delta^{-1}\boldsymbol{\Lambda}$.

3.1 NIG Expression for Prior Distribution

Recall the likelihood function (1) of quantile regression and denote $\hat{\boldsymbol{\beta}}_p = (\mathbf{X}^{*T}\boldsymbol{\Sigma}^{-1}\mathbf{X}^*)^{-1}\mathbf{X}^{*T}\boldsymbol{\Sigma}^{-1}\mathbf{Y}_p^*$, we can rewrite likelihood (1) as

$$
\begin{aligned}
f(\mathbf{Y}_p^*|\boldsymbol{\beta},\sigma,\mathbf{v},\mathbf{X}^*) &\propto \sigma^{-\frac{n-k}{2}}\exp\{-\frac{1}{2\sigma}[\mathbf{Y}_p^*-\mathbf{X}^*\hat{\boldsymbol{\beta}}_p]^T\boldsymbol{\Sigma}^{-1}[\mathbf{Y}_p^*-\mathbf{X}^*\hat{\boldsymbol{\beta}}_p]\} \\
&\quad \sigma^{-\frac{k}{2}}\exp\{-\frac{1}{2\sigma}(\boldsymbol{\beta}-\hat{\boldsymbol{\beta}}_p)^T(\mathbf{X}^{*T}\boldsymbol{\Sigma}^{-1}\mathbf{X}^*)(\boldsymbol{\beta}-\hat{\boldsymbol{\beta}}_p)\} \\
&= (\sigma)^{-(a+\frac{k}{2}+1)}\exp\{-\frac{1}{\sigma}[b_p+\frac{1}{2}(\boldsymbol{\beta}-\boldsymbol{\mu}_p)^T\boldsymbol{\Lambda}(\boldsymbol{\beta}-\boldsymbol{\mu}_p)]\} \\
&\propto IG(a,b_p)N_k(\boldsymbol{\mu}_p,\sigma\boldsymbol{\Lambda}^{-1}),
\end{aligned} \tag{2}
$$

where $\boldsymbol{\mu}_p=\hat{\boldsymbol{\beta}}_p$, $\boldsymbol{\Lambda}=\mathbf{X}^{*T}\boldsymbol{\Sigma}^{-1}\mathbf{X}^*$, $a=\frac{n-k-2}{2}$ and $b_p=\frac{1}{2}[\mathbf{Y}_p^*-\mathbf{X}^*\hat{\boldsymbol{\beta}}_p]^T\boldsymbol{\Sigma}^{-1}[\mathbf{Y}_p^*-\mathbf{X}^*\hat{\boldsymbol{\beta}}_p]$. According to Definition 1 with $\delta=\sigma$, the rewritten likelihood (2) can be represented as the structure of a k-dimensional distribution $NIG_k(\boldsymbol{\mu},\boldsymbol{\Lambda},a,b)$ in terms of parameters $(\boldsymbol{\beta},\sigma)$.

Under the informative prior setting, following Alhamzawi and Yu [1], a conjugate prior for $(\boldsymbol{\beta},\sigma)$ with a modification of Zellner's informative g-prior [28] in QR could be provided as

$$
\boldsymbol{\beta}|\sigma,\mathbf{v},\mathbf{X}^*,\boldsymbol{\Sigma}\sim N_k(\mathbf{0}_k,g\sigma(\mathbf{X}^{*T}\boldsymbol{\Sigma}^{-1}\mathbf{X}^*)^{-1}),\ f(\sigma)\propto\sigma^{-1},
$$

where $g>0$ is a known scaling factor prescribed by the user. Smith and Kohn [20] proposed a Bayesian variable selection algorithm utilizing regression splines. They found that the choice of $g=100$ works well and suggested to choose g between 10 and 1000. Following Smith and Kohn [20], the fixed setting of $g=100$ has been considered by some other authors (see [8, 13], among others). Then we obtain the joint prior distribution of $(\boldsymbol{\beta},\sigma)$

$$
f(\boldsymbol{\beta},\sigma|\mathbf{v},\mathbf{X}^*,\boldsymbol{\Sigma})\propto\sigma^{-(\frac{k}{2}+1)}\exp\{-\frac{1}{\sigma}[\frac{1}{2}\boldsymbol{\beta}^T\frac{\mathbf{X}^{*T}\boldsymbol{\Sigma}^{-1}\mathbf{X}^*}{g}\boldsymbol{\beta}]\}, \tag{3}
$$

which is a special case of $NIG_k(\boldsymbol{\mu}_0,\boldsymbol{\Lambda}_{g0},a_0,b_0)$ with $\boldsymbol{\mu}_0=\mathbf{0}_k$, $\boldsymbol{\Lambda}_{g0}=\frac{\mathbf{X}^{*T}\boldsymbol{\Sigma}^{-1}\mathbf{X}^*}{g}$, $a_0=0$, $b_0=0$.

3.2 *NIG Expression for Posterior Distribution*

The joint conditional posterior distribution $f(\boldsymbol{\beta},\sigma,\mathbf{v}|\mathbf{Y}_p^*,\mathbf{X}^*)$ under the informative g-prior (3) is given by

$$
\begin{aligned}
f(\boldsymbol{\beta},\sigma,\mathbf{v}|\mathbf{Y}_p^*,\mathbf{X}^*) &\propto f(\mathbf{Y}_p^*|\boldsymbol{\beta},\sigma,\mathbf{v})f(\boldsymbol{\beta}|\sigma,\mathbf{v})f(\mathbf{v}|\sigma)f(\sigma)\\
&\propto \sigma^{-(\frac{3n+k+2}{2})}(\prod_{i=1}^{n} v_i^{-1/2})|\mathbf{X}^{*T}\boldsymbol{\Sigma}^{-1}\mathbf{X}^*|^{1/2}\\
&\times \exp\{-\frac{1}{2\sigma}[(\mathbf{Y}_p^*-\mathbf{X}^*\boldsymbol{\beta})^T\boldsymbol{\Sigma}^{-1}(\mathbf{Y}_p^*-\mathbf{X}^*\boldsymbol{\beta})\\
&+\boldsymbol{\beta}^T\frac{\mathbf{X}^{*T}\boldsymbol{\Sigma}^{-1}\mathbf{X}^*}{g}\boldsymbol{\beta}+2p(1-p)\sum_{i=1}^{n} v_i]\}.
\end{aligned}
$$

Then the corresponding posterior $f(\boldsymbol{\beta},\sigma|\mathbf{v},\mathbf{Y}_p^*,\mathbf{X}^*)$ is given as follows:

$$
\begin{aligned}
f(\boldsymbol{\beta},\sigma|\mathbf{v},\mathbf{Y}_p^*,\mathbf{X}^*) &\propto \sigma^{-(\frac{3n+k+2}{2})}\exp\{-\frac{1}{2\sigma}[(\mathbf{Y}_p^*-\mathbf{X}^*\boldsymbol{\beta})^T\boldsymbol{\Sigma}^{-1}(\mathbf{Y}_p^*-\mathbf{X}^*\boldsymbol{\beta})\\
&+\boldsymbol{\beta}^T\frac{\mathbf{X}^{*T}\boldsymbol{\Sigma}^{-1}\mathbf{X}^*}{g}\boldsymbol{\beta}+2p(1-p)\sum_{i=1}^{n} v_i]\}\\
&=\sigma^{-(\frac{3n}{2}+\frac{k}{2}+1)}\exp\{-\frac{1}{\sigma}[\bar{b}_p+\frac{1}{2}(\boldsymbol{\beta}-\bar{\boldsymbol{\mu}}_p)^T\bar{\boldsymbol{\Lambda}}(\boldsymbol{\beta}-\bar{\boldsymbol{\mu}}_p)]\},
\end{aligned}
$$

which has an expression of $NIG_k(\bar{\boldsymbol{\mu}}_p,\bar{\boldsymbol{\Lambda}},\bar{a},\bar{b}_p)$, where $\bar{\boldsymbol{\mu}}_p=[(1+\frac{1}{g})\mathbf{X}^{*T}\boldsymbol{\Sigma}^{-1}\mathbf{X}^*]^{-1}\mathbf{X}^{*T}\boldsymbol{\Sigma}^{-1}\mathbf{Y}_p^*$, $\bar{\boldsymbol{\Lambda}}=(1+\frac{1}{g})\mathbf{X}^{*T}\boldsymbol{\Sigma}^{-1}\mathbf{X}^*$, $\bar{a}=\frac{3n}{2}$, $\bar{b}_p=\frac{1}{2}\mathbf{Y}_p^{*T}\boldsymbol{\Sigma}^{-1}\mathbf{Y}_p^*-\frac{1}{2}\bar{\boldsymbol{\mu}}_p^T\bar{\boldsymbol{\Lambda}}\bar{\boldsymbol{\mu}}_p+p(1-p)\sum_{i=1}^n v_i$. Moreover, the full conditional distributions of $\boldsymbol{\beta}$ and σ can be obtained respectively by

$$
f(\boldsymbol{\beta}|\sigma,\mathbf{v},\mathbf{Y}_p^*,\mathbf{X}^*)\propto\exp\{-\frac{1}{2\sigma}[(\mathbf{Y}_p^*-\mathbf{X}^*\boldsymbol{\beta})^T\boldsymbol{\Sigma}^{-1}(\mathbf{Y}_p^*-\mathbf{X}^*\boldsymbol{\beta})+\boldsymbol{\beta}^T\frac{\mathbf{X}^{*T}\boldsymbol{\Sigma}^{-1}\mathbf{X}^*}{g}\boldsymbol{\beta}]\},
$$

which can be expressed as a k-dimensional normal $N_k(\bar{\boldsymbol{\mu}}_p,\sigma\bar{\boldsymbol{\Lambda}}^{-1})$, and

$$
\begin{aligned}
f(\sigma|\boldsymbol{\beta},\mathbf{v},\mathbf{Y}_p^*,\mathbf{X}^*) &\propto \sigma^{-(\frac{3n+k}{2}+1)}\exp\{-\frac{1}{2\sigma}[(\mathbf{Y}_p^*-\mathbf{X}^*\boldsymbol{\beta})^T\boldsymbol{\Sigma}^{-1}(\mathbf{Y}_p^*-\mathbf{X}^*\boldsymbol{\beta})\\
&+\boldsymbol{\beta}^T\frac{\mathbf{X}^{*T}\boldsymbol{\Sigma}^{-1}\mathbf{X}^*}{g}\boldsymbol{\beta}+2p(1-p)\sum_{i=1}^{n} v_i]\},
\end{aligned}
$$

which is an IG distribution with shape $\frac{3n+k}{2}$ and scale $\frac{1}{2}[(\mathbf{Y}_p^*-\mathbf{X}^*\boldsymbol{\beta})^T\boldsymbol{\Sigma}^{-1}(\mathbf{Y}_p^*-\mathbf{X}^*\boldsymbol{\beta})+\boldsymbol{\beta}^T\frac{\mathbf{X}^{*T}\boldsymbol{\Sigma}^{-1}\mathbf{X}^*}{g}\boldsymbol{\beta}+2p(1-p)\sum_{i=1}^n v_i]$. The full posterior distribution of each v_i, $i=1,2,\ldots,n$ is also tractable:

$$
\begin{aligned}
f(v_i|\boldsymbol{\beta},\sigma,y_i,\mathbf{x}_i) &\propto v_i^{-1}\exp\{-\frac{1}{4\sigma}[vi^{-1}((y_i-(1-2p)v_i-\mathbf{x}_i^T\boldsymbol{\beta})^2+\frac{\boldsymbol{\beta}^T\mathbf{x}_i\mathbf{x}_i^T\boldsymbol{\beta}}{g})]-\frac{p(1-p)}{\sigma}v_i\}\\
&=v_i^{-1}\exp\{-\frac{1}{4\sigma}[v_i^{-1}((y_i-\mathbf{x}_i^T\boldsymbol{\beta})^2+\frac{\boldsymbol{\beta}^T\mathbf{x}_i\mathbf{x}_i^T\boldsymbol{\beta}}{g})+v_i]\}\\
&=v_i^{-1}\exp\{-\frac{1}{2}(v_i^{-1}\bar{\xi}_i^2+v_i\bar{\zeta}_i^2)\},
\end{aligned}
$$

where $\bar{\xi}_i^2 = [(y_i - \mathbf{x}_i^T\boldsymbol{\beta})^2 + \boldsymbol{\beta}^T\mathbf{x}_i\mathbf{x}_i^T\boldsymbol{\beta}/g]/2\sigma$ and $\bar{\zeta}_i^2 = 1/2\sigma$, which can be recognized as a generalized inverse Gaussian distribution $GIG(0, \bar{\xi}_i, \bar{\zeta}_i)$ [2].

4 Big Data Based Algorithms for Bayesian Quantile Regression

4.1 NIG Multiplication Operator for Posterior Distribution

To derive the posterior distribution induced by the entire data set for Bayesian quantile regression, we first introduce the NIG multiplication operator defined as follows.

Proposition 1 *A general k-dimensional normal-inverse-gamma distribution $NIG_k(\boldsymbol{\mu}, \boldsymbol{\Lambda}, a, b)$ can be reformulated as a multiplication of H independent k-dimensional distributions $NIG_k(\boldsymbol{\mu}_h, \boldsymbol{\Lambda}_h, a_h, b_h), h = 1, \ldots, H$*

$$NIG_k(\boldsymbol{\mu}, \boldsymbol{\Lambda}, a, b) = \prod_{h=1}^{H} NIG_k(\boldsymbol{\mu}_h, \boldsymbol{\Lambda}_h, a_h, b_h), \tag{4}$$

where $\boldsymbol{\mu} = (\sum_{h=1}^H \boldsymbol{\Lambda}_h)^{-1}\sum_{h=1}^H \boldsymbol{\Lambda}_h\boldsymbol{\mu}_h$, $\boldsymbol{\Lambda} = \sum_{h=1}^H \boldsymbol{\Lambda}_h$, $a = \sum_{h=1}^H a_h + \frac{(H-1)(k+2)}{2}$ *and* $b = \sum_{h=1}^H b_h + \frac{1}{2}\sum_{h=1}^H(\boldsymbol{\mu}_h - \boldsymbol{\mu})^T\boldsymbol{\Lambda}_h(\boldsymbol{\mu}_h - \boldsymbol{\mu})$.

Recall the rewritten likelihood function of quantile regression (2) given in Sect. 3.1. If we partition the big data of $\mathbf{X}^*$ and $\mathbf{Y}_p^*$ into M subsets, where each $\mathbf{X}_m^*$ is an $n_m \times k$ matrix, $\mathbf{Y}_{pm}^*$ is an $n_m \times 1$ vector, $\boldsymbol{\Sigma}_m$ is an $n_m \times n_m$ diagonal block of $\boldsymbol{\Sigma}$ and $\sum_{m=1}^M n_m = n$, then the likelihood (2) can be reformulated as

$$f(\mathbf{Y}_p^*|\boldsymbol{\beta}, \sigma, \mathbf{v}, \mathbf{X}^*) \propto \sigma^{-\frac{\sum_{m=1}^M n_m - k}{2}} \exp\{-\frac{1}{2\sigma}\sum_{m=1}^M[\mathbf{Y}_{pm}^* - \mathbf{X}_m^*\hat{\boldsymbol{\beta}}_p]^T\boldsymbol{\Sigma}_m^{-1}[\mathbf{Y}_{pm}^* - \mathbf{X}_m^*\hat{\boldsymbol{\beta}}_p]\}$$
$$\sigma^{-\frac{k}{2}}\exp\{-\frac{1}{2\sigma}\sum_{m=1}^M(\boldsymbol{\beta} - \hat{\boldsymbol{\beta}}_p)^T(\mathbf{X}_m^{*T}\boldsymbol{\Sigma}_m^{-1}\mathbf{X}_m^*)(\boldsymbol{\beta} - \hat{\boldsymbol{\beta}}_p)\},$$

which indicates a multiplication of M NIG distributions regarding parameters $(\boldsymbol{\beta}, \sigma)$

$$f(\mathbf{Y}_p^*|\boldsymbol{\beta}, \sigma, \mathbf{v}, \mathbf{X}^*) \propto \prod_{m=1}^M \sigma^{-(a_m^{(l)} + \frac{k}{2} + 1)}\exp\{-\frac{1}{\sigma}[b_m^{(l)} + \frac{1}{2}(\boldsymbol{\beta} - \boldsymbol{\mu}_p^{(l)})^T\boldsymbol{\Lambda}_m(\boldsymbol{\beta} - \boldsymbol{\mu}_p^{(l)})]\}$$
$$= \prod_{m=1}^M NIG(\boldsymbol{\mu}_p^{(l)}, \boldsymbol{\Lambda}_m^{(l)}, a_m^{(l)}, b_{pm}^{(l)}),$$

where the superscript (l) indicates the NIG parameters concerning $(\boldsymbol{\beta}, \sigma)$ for the likelihood function. $\boldsymbol{\mu}_p^{(l)} = \hat{\boldsymbol{\beta}}_p = (\sum_{m=1}^M \mathbf{X}_m^{*T}\boldsymbol{\Sigma}_m^{-1}\mathbf{X}_m^*)^{-1}\sum_{m=1}^M \mathbf{X}_m^{*T}\boldsymbol{\Sigma}_m^{-1}\mathbf{Y}_{pm}^*$, $\boldsymbol{\Lambda}_m^{(l)} =$

$\mathbf{X}_m^{*T}\boldsymbol{\Sigma}_m^{-1}\mathbf{X}_m^*$, $a_m^{(l)} = \frac{n_m-k-2}{2}$ and $b_{pm}^{(l)} = \frac{1}{2}[\mathbf{Y}_{pm}^* - \mathbf{X}_m^*\boldsymbol{\mu}_p^{(l)}]^T\boldsymbol{\Sigma}_m^{-1}[\mathbf{Y}_{pm}^* - \mathbf{X}_m^*\boldsymbol{\mu}_p^{(l)}]$. Then the full data posterior distribution is calibrated by the product of specified NIG prior and this multiplicative likelihood function, employing Eq. (4) with $H = M + 1$ in this case. The following Theorem 1 elaborates the acquisition of posterior distribution through the use of NIG multiplication operators.

Theorem 1 *Consider a linear quantile regression model with full big data observations* $\mathbf{X}$ *and* $\mathbf{Y}$. *Denote the posterior distribution of regression parameters* $(\boldsymbol{\beta}, \sigma)$, *under the prior* $NIG_k(\boldsymbol{\mu}^{(0)}, \boldsymbol{\Lambda}^{(0)}, a^{(0)}, b^{(0)})$, *be* $NIG_k(\bar{\boldsymbol{\mu}}, \bar{\boldsymbol{\Lambda}}, \bar{a}, \bar{b})$. *If we partition the whole data of size* n *into* M *subsets, each with an* $n_m \times k$ *matrix* $\mathbf{X}_m$ *and an* $n_m \times 1$ *vector* $\mathbf{Y}_m$, $m = 1, \ldots, M$, *and let* $\mathbf{X}_m^* = \frac{1}{\sqrt{2}}\mathbf{X}_m$, $\mathbf{Y}_{pm}^* = \frac{1}{\sqrt{2}}(\mathbf{Y}_m - (1 - 2p)\mathbf{v}_m)$, $\boldsymbol{\Sigma}_m = \text{diag}(\mathbf{v}_m)$, *where the latent variable* $\mathbf{v}_m$ *is an* $n_m \times 1$ *vector generated from the exponential distribution with rate* $\sigma^{-1}p(1-p)$, *then the full data posterior distribution can be formulated as*

$$NIG_k(\bar{\boldsymbol{\mu}}, \bar{\boldsymbol{\Lambda}}, \bar{a}, \bar{b}) = NIG_k(\boldsymbol{\mu}^{(0)}, \boldsymbol{\Lambda}^{(0)}, a^{(0)}, b^{(0)}) \prod_{m=1}^{M} NIG_k(\boldsymbol{\mu}_p^{(l)}, \boldsymbol{\Lambda}_m^{(l)}, a_m^{(l)}, b_{pm}^{(l)}),$$

where $\bar{\boldsymbol{\mu}} = (\boldsymbol{\Lambda}^{(0)} + \sum_{m=1}^{M}\mathbf{X}_m^{*T}\boldsymbol{\Sigma}_m^{-1}\mathbf{X}_m^*)^{-1}(\boldsymbol{\Lambda}^{(0)}\boldsymbol{\mu}^{(0)} + \sum_{m=1}^{M}\mathbf{X}_m^{*T}\boldsymbol{\Sigma}_m^{-1}\mathbf{Y}_{pm}^*)$, $\bar{\boldsymbol{\Lambda}} = \boldsymbol{\Lambda}^{(0)} + \sum_{m=1}^{M}\mathbf{X}_m^{*T}\boldsymbol{\Sigma}_m^{-1}\mathbf{X}_m^*$, $\bar{a} = a^{(0)} + \frac{n}{2}$ *and* $\bar{b} = b^{(0)} + \frac{1}{2}[\sum_{m=1}^{M}\mathbf{Y}_{pm}^{*T}\boldsymbol{\Sigma}_m^{-1}\mathbf{Y}_{pm}^* + \boldsymbol{\mu}^{(0)T}\boldsymbol{\Lambda}^{(0)}\boldsymbol{\mu}^{(0)} - \bar{\boldsymbol{\mu}}^T\bar{\boldsymbol{\Lambda}}\bar{\boldsymbol{\mu}}]$.

4.2 Algorithms for Bayesian Quantile Regression

Consider the linear QR model for the p-th quantile $(0 < p < 1)$

$$\mathbf{Y} = \mathbf{X}\boldsymbol{\beta} + \boldsymbol{\varepsilon}, \tag{5}$$

where $\mathbf{Y}$ is an $n \times 1$ response vector, $\mathbf{X}$ is an $n \times k$ predictor matrix, and $\boldsymbol{\varepsilon}$ is an $n \times 1$ vector of $ALD(0, \sigma, p)$ disturbances. Then model (5) is equivalent to

$$\mathbf{Y}_p^* = \mathbf{X}^*\boldsymbol{\beta} + \sqrt{\sigma}\boldsymbol{\varepsilon}^*,$$

where $\mathbf{Y}_p^* = \frac{1}{\sqrt{2}}(\mathbf{Y} - (1-2p)\mathbf{v})$, $\mathbf{X}^* = \frac{1}{\sqrt{2}}\mathbf{X}$ and $\boldsymbol{\varepsilon}^* \sim N_n(\mathbf{0}_n, \boldsymbol{\Sigma})$ with $n \times n$ known positive definite covariance matrix $\boldsymbol{\Sigma}$. Then we proceed to Bayesian inference for big data quantile regressions through the proposed NIG multiplication operator. We consider model (5) under the g-prior (3) for $(\boldsymbol{\beta}, \sigma)$, and partition the entire data set into M subsets $(\mathbf{X}_m, \mathbf{Y}_m)$ with individual sample size n_m, $m = 1, \ldots, M$. Then the posterior distribution for the whole data can be obtained by merging the given prior with the multiplication of M subset NIG distributions induced from the massive observations. Based on this, an efficient divide-and-conquer algorithm for big data Bayesian quantile regression is provided as below.

Algorithm 1 Consider a pth $(0 < p < 1)$ Bayesian quantile regression under g-prior (3) with the observed $n \times k$ design matrix $\mathbf{X}$ and $n \times 1$ response vector $\mathbf{Y}$, where the large data set cannot be fit into a single computer due to the memory constraint. We can obtain the full data posterior distribution by the following divide-and-conquer algorithm.

Step 1 partition the entire data set into M subsets $\mathbf{X}_m, \mathbf{Y}_m, m = 1, 2, \ldots, M$, where $\mathbf{X}_m$ is an $n_m \times k$ matrix, $\mathbf{Y}_m$ is an $n_m \times 1$ vector and $\sum_{m=1}^{M} n_m = n$.

Step 2 for each subset $\mathbf{X}_m, \mathbf{Y}_m$, a Gibbs sampler for sampling $\boldsymbol{\beta}_m, \sigma_m$ and $\mathbf{v}_m$ in the BQR would follow the sub-steps presented below:

2.1 denote j as the iteration count. Then set $j = 0$ and establish $(\boldsymbol{\beta}_m^{(j=0)}, \sigma_m^{(j=0)}, \mathbf{v}_m^{(j=0)})$ to some starting values.

2.2 follow the full conditional distributions of $\boldsymbol{\beta}_m, \sigma_m$ and $\mathbf{v}_m$,

(i) sample $\mathbf{v}_m^{(j+1)}$ from $f(\mathbf{v}_m|\boldsymbol{\beta}_m^{(0)}, \sigma_m^{(0)})$.

(ii) sample $\sigma_m^{(j+1)}$ from $f(\sigma_m|\boldsymbol{\beta}_m^{(0)}, \mathbf{v}_m^{(1)})$.

(iii) sample $\boldsymbol{\beta}_m^{(j+1)}$ from $f(\boldsymbol{\beta}_m|\sigma_m^{(1)}, \mathbf{v}_m^{(1)})$.

2.3 set $j = j + 1$ and return to **Step 2.2** until $j = L$, where L is the number of iteration times.

Step 3 calculate the empirical estimates of the means $\bar{\boldsymbol{\beta}}_m$ and $\bar{\sigma}_m$ separately based on the $(L - B)$ realizations of the Gibbs sequence (discarding the first B iterations as a burn-in). Then generate an n_m i.i.d. sample on υ_i, where $\upsilon_i \sim GIG(0, \bar{\xi}_i, \bar{\zeta}_i)$, with $\bar{\xi}_i^2 = [(y_i - \mathbf{x}_i^T\bar{\boldsymbol{\beta}}_m)^2 + \bar{\boldsymbol{\beta}}_m^T\mathbf{x}_i\mathbf{x}_i^T\bar{\boldsymbol{\beta}}_m/g]/2\bar{\sigma}_m$ and $\bar{\zeta}_i^2 = 1/2\bar{\sigma}_m, i = 1, 2, \ldots, n_m$. Let $\mathbf{X}_m^* = \frac{1}{\sqrt{2}}\mathbf{X}_m$, $\mathbf{Y}_{pm}^* = \frac{1}{\sqrt{2}}(\mathbf{Y}_m - (1 - 2p)\mathbf{v}_m)$, where $\mathbf{v}_m$ is the corresponding $n_m \times 1$ vector of υ_i for each subset, and denote $\boldsymbol{\Sigma}_m$ as an $n_m \times n_m$ diagonal matrix with $\mathbf{v}_m$ its diagonal vector, $m = 1, 2, \ldots, M$.

Step 4 for each subset, the corresponding likelihood can be represented as a form of $NIG_k(\boldsymbol{\mu}_{pm}, \boldsymbol{\Lambda}_m, a_m, b_{pm})$ distribution for $(\boldsymbol{\beta}, \sigma)$. Obtain the multiplicative distribution $NIG_k(\boldsymbol{\mu}_p, \boldsymbol{\Lambda}, a, b_p) = \prod_{m=1}^{M} NIG(\boldsymbol{\mu}_{pm}, \boldsymbol{\Lambda}_m, a_m, b_{pm})$, then the full data posterior can be given by merging the g-prior $NIG_k(\boldsymbol{\mu}_0, \boldsymbol{\Lambda}_{g0}, a_0, b_0)$ and distribution $NIG_k(\boldsymbol{\mu}_p, \boldsymbol{\Lambda}, a, b_p)$:

$$NIG_k(\bar{\boldsymbol{\mu}}_p, \bar{\boldsymbol{\Lambda}}, \bar{a}, \bar{b}_p) = NIG_k(\boldsymbol{\mu}_0, \boldsymbol{\Lambda}_{g0}, a_0, b_0)NIG_k(\boldsymbol{\mu}_p, \boldsymbol{\Lambda}, a, b_p),$$

where $\bar{\boldsymbol{\mu}}_p = [(1 + \frac{1}{g})\sum_{m=1}^{M} \mathbf{X}_m^{*T}\boldsymbol{\Sigma}_m^{-1}\mathbf{X}_m^*)]^{-1}\sum_{m=1}^{M} \mathbf{X}_m^{*T}\boldsymbol{\Sigma}_m^{-1}\mathbf{Y}_{pm}^*$, $\bar{\boldsymbol{\Lambda}} = (1 + \frac{1}{g})\sum_{m=1}^{M} \mathbf{X}_m^{*T}\boldsymbol{\Sigma}_m^{-1}\mathbf{X}_m^*$, $\bar{a} = \frac{3n}{2}$, $\bar{b}_p = \frac{1}{2}[\sum_{m=1}^{M} \mathbf{Y}_{pm}^{*T}\boldsymbol{\Sigma}_m^{-1}\mathbf{Y}_{pm}^* - \bar{\boldsymbol{\mu}}_p^T\bar{\boldsymbol{\Lambda}}\bar{\boldsymbol{\mu}}_p] + p(1 - p)\sum_{m=1}^{M} \|\mathbf{v_m}\|_1$ and $\|\cdot\|_1$ denotes the ℓ_1 norm of a vector.

Table 1 Summary statistics for wind power observations at Aeolos, Iweco and Rokas

	Aeolos	Iweco	Rokas
Min	0.000	0.000	0.000
Quantile (0.25)	1.692	0.921	1.573
Median	4.002	2.112	4.579
Mean	4.142	2.141	4.857
Quantile (0.75)	6.745	3.426	8.049
Max	8.302	4.549	11.635
Standard deviation	2.649	1.346	3.407
Sample size	17,819	15,621	21,949

5 Real-Data Analysis

In this section, we illustrate our divide-and-conquer algorithm for big data Bayesian quantile regression by a real-world data analysis. We use hourly wind power data recorded from 31 December 2007 to 30 December 2010 at the following three wind farms in Crete: Aeolos, Iweco and Rokas. The data is a collection of hourly observations for wind speed (measured in m/s), direction (measured in degrees) and power (measured in megawatts). A complete wind power data of the year 2010 is examined in Taylor [21]. We remove all the missing data and retain positive observations of the recorded hourly periods. Table 1 presents the summary statistics for wind power observations (in MW) at Aeolos, Iweco and Rokas respectively.

We fit our big data BQR by modeling the wind power as a linear function of wind speed and direction. We implement Algorithm 1 for these three power sequences at $p = 0.50$ and $p = 0.95$ respectively. In each case, the Gibbs samplers are run for 11000 iterations, discarding the first 1000 as a burn-in. For Aeolos farm, the whole observations are partitioned into 50 subsets with the size of $n_1 = n_2 \ldots = n_{49} = 356$ and $n_{50} = 375$. For Iweco, we partition the whole data into 50 subsets with the size of $n_1 = n_2 \ldots = n_{49} = 312$ and $n_{50} = 333$. For Rokas, we consider 50 subsets as $n_1 = n_2 \ldots = n_{49} = 438$ and $n_{50} = 487$. We assign the informative g-prior by choosing $g = 100$. Table 2 displays the estimates and posterior standard deviations in our big data BQR model for the given three wind power series separately. Note that for all power series, the estimated coefficients of direction are close to zero at the measured percentiles, meaning that the effect of wind direction on power seems to be minor. Instead, wind power presents a much stronger correlation to speed than to direction. The positive coefficients of speed indicate that as wind speed increases, so does the power capacity. Furthermore, it is visible that speed has a greater impact on higher (95th percentile) power than lower (50th percentile) power capacity for all the three aforementioned wind farms.

Table 2 Coefficient estimates along with posterior standard deviations (S.D.) for Aeolos, Iweco and Rokas in big data BQR analysis

Model covariates	Aeolos				Iweco				Rokas			
	$p = 0.50$		$p = 0.95$		$p = 0.50$		$p = 0.95$		$p = 0.50$		$p = 0.95$	
	Coeff.	S.D.	Coeff.	S.D.	Coeff.	S.D.	Coeff.	S.D.	Coeff.	S.D.	Coeff.	S.D.
Intercept	−2.8624	0.0151	−3.6681	0.0160	−0.7907	0.0110	−0.5663	0.0135	−2.8004	0.0130	−1.9270	0.0150
Speed	0.7485	0.0151	1.0447	0.0018	0.3770	0.0015	0.5316	0.0015	0.7860	0.0013	1.0616	0.0010
Direction	−0.0003	0.0000	−0.0021	0.0000	−0.0023	0.0000	−0.0039	0.0000	0.0005	0.0000	−0.0040	0.0000

6 Summary and Conclusion

This paper extends the divide-and-conquer algorithm for big data analysis from traditional mean-based linear regression to quantile regression under Bayesian perspectives. This is achieved by using ALD-based working likelihood functions and conjugate NIG priors. The resulting algorithms are easily implemented and the real-data illustrations present that wind speed has a greater impact on higher power values than lower ones, showing the proposed methods are promising. The developed algorithms can be investigated for other energy-related observations within big data scenario, such as solar radiation and electrical power demand series. In this empirical study, we have assigned the positive scaling g-prior by fixing it to be the experimental value $g = 100$, as suggested in Smith and Kohn [20] after extensive testing. However, a potential alternative is to assign a hyper-prior distribution on the g parameter rather than keep it as a fixed constant. Under such circumstances, the unknown parameter g can be estimated from the available data. Moreover, the undesirable "Information Paradox", which relates to the limiting behavior of the Bayes factor for model selection with fixed g, can be avoided (see [14, 16]). Our possible future work will focus on developing a novel Bayesian quantile regression for fitting single-index models under high-dimensional data context, and its penalized version for efficient variable selection implementations.

Acknowledgements The authors would like to thank the support of the National Social Science Foundation of China (Series number: 21BTJ040).

References

1. Alhamzawi, R., Yu, K.: Conjugate priors and variable selection for Bayesian quantile regression. Comput. Stat. Data Anal. **64**, 209–219 (2013)
2. Barndorff-Nielsen, O.E., Shephard, N.: Non-Gaussian Ornstein-Uhlenbeck-based models and some of their uses in financial economics. J. Roy. Stat. Soc. B: Stat. Methodol. **63**, 167–241 (2001)
3. Briollais, L., Durrieu, G.: Application of quantile regression to recent genetic and -omic studies. Hum. Genet. **133**, 951–966 (2014)
4. Bernardi, M., Gayraud, G., Petrella, L.: Bayesian tail risk interdependence using quantile regression. Bayesian Anal. **10**, 553–603 (2015)
5. Cole, T.J., Green, P.J.: Smoothing reference centile curves: the LMS method and penalized likelihood. Stat. Med. **11**, 1305–1319 (1992)
6. Chen, X., Liu, W., Zhang, Y.: Quantile regression under memory constraint. Ann. Stat. **47**, 3244–3273 (2019)
7. Denison, D.G., Holmes, C.C., Mallick, B.K., Smith, A.F.: Bayesian Methods for Nonlinear Classification and Regression. Wiley, Hoboken (2002)
8. Gupta, M., Qu, P., Ibrahim, J.G.: A temporal hidden Markov regression model for the analysis of gene regulatory networks. Biostatistics **8**, 805–820 (2007)
9. Gu, Y., Fan, J., Kong, L., Ma, S., Zou, H.: ADMM for high-dimensional sparse penalized quantile regression. Technometrics **60**, 319–331 (2018)
10. Gonçalves, K.C., Migon, H.S., Bastos, L.S.: Dynamic quantile linear models: a Bayesian approach. Bayesian Anal. **15**, 335–362 (2020)

11. Koenker, R., Hallock, K.F.: Quantile regression: an introduction. J. Econ. Perspect. **15**, 143–156 (2001)
12. Kozumi, H., Kobayashi, G.: Gibbs sampling methods for Bayesian quantile regression. J. Stat. Comput. Simul. **81**, 1565–1578 (2011)
13. Lee, K.E., Sha, N., Dougherty, E.R., Vannucci, M., Mallick, B.K.: Gene selection: a Bayesian variable selection approach. Bioinformatics **19**, 90–97 (2003)
14. Liang, F., Paulo, R., Molina, G., Clyde, M.A., Berger, J.O.: Mixtures of g priors for Bayesian variable selection. J. Am. Stat. Assoc. **103**, 410–423 (2008)
15. Lum, K., Gelfand, A.E.: Spatial quantile multiple regression using the asymmetric Laplace process. Bayesian Anal. **7**, 235–258 (2012)
16. Perrakis, K., Ntzoufras, I.: Bayesian variable selection using the hyper-g prior in WinBUGS. Wiley Interdisc. Rev. Comput. Stat. **10**, e1442 (2018)
17. Petrella, L., Raponi, V.: Joint estimation of conditional quantiles in multivariate linear regression models with an application to financial distress. J. Multivar. Anal. **173**, 70–84 (2019)
18. Reed, C., Yu, K.: A partially collapsed Gibbs sampler for Bayesian quantile regression (2009)
19. Rodrigues, T., Fan, Y.: Regression adjustment for noncrossing Bayesian quantile regression. J. Comput. Graph. Stat. **26**, 275–284 (2017)
20. Smith, M., Kohn, R.: Nonparametric regression using Bayesian variable selection. J. Econom. **75**, 317–343 (1996)
21. Taylor, J.W.: Probabilistic forecasting of wind power ramp events using autoregressive logit models. Eur. J. Oper. Res. **259**, 703–712 (2017)
22. Wu, Y., Yin, G.: Conditional quantile screening in ultrahigh-dimensional heterogeneous data. Biometrika **102**, 65–76 (2015)
23. Wang, Y., Feng, X.N., Song, X.Y.: Bayesian quantile structural equation models. Struct. Equ. Model. **23**, 246–258 (2016)
24. Yu, K., Moyeed, R.A.: Bayesian quantile regression. Stat. Probab. Lett. **54**, 437–447 (2001)
25. Yu, K., Lu, Z., Stander, J.: Quantile regression: applications and current research areas. J. Roy. Stat. Soc. Ser. D Stat. **52**, 331–350 (2003)
26. Yu, K., Stander, J.: Bayesian analysis of a Tobit quantile regression model. J. Econ. **137**, 260–276 (2007)
27. Yu, L., Lin, N., Wang, L.: A parallel algorithm for large-scale nonconvex penalized quantile regression. J. Comput. Graph. Stat. **26**, 935–939 (2017)
28. Zellner, A.: On assessing prior distributions and Bayesian regression analysis with g-prior distributions. In: Goel P.K., Zellner, A. (eds.) Bayesian Inference and Decision Techniques: Essays in Honor of Bruno de Finetti, pp. 233–243. Elsevier, North-Holland (1986)

Towards a Bayesian Analysis of Migration Pathways Using Chain Event Graphs of Agent Based Models

Peter Strong, Alys McAlpine, and Jim Q. Smith

Abstract Agent-Based Models (ABMs) are often used to model migration and are increasingly used to simulate individual migrant decision-making and unfolding events through a sequence of heuristic if-then rules. However, ABMs lack the methods to embed more principled strategies of performing inference to estimate and validate the models, both of which are of significant importance for real-world case studies. Chain Event Graphs (CEGs) can fill this need: they can be used to provide a Bayesian framework which represents an ABM accurately. Through the use of the CEG, we illustrate how to transform an elicited ABM into a Bayesian framework and outline the benefits of this approach.

Keywords Applied statistics · Probabilistic graphical models · Context-specific independence · Conditional independence

1 Introduction

Researchers and policymakers are interested in modelling migration as they aim to understand the mechanisms involved in order to inform policy. For example, organisations may aim to promote safe labour migration in line with the UN's Sustainable Development Goals [22]. Migration can increase vulnerability to human traffick-

P. Strong (✉)
Centre for Complexity Science, University of Warwick, Coventry CV4 7AL, UK
e-mail: P.R.Strong@warwick.ac.uk

P. Strong · J. Q. Smith
The Alan Turing Institute, British Library, 96 Euston Road, London NW1 2DB, UK
e-mail: J.Q.Smith@warwick.ac.uk

A. McAlpine
Gender Violence and Health Centre, Faculty of Public Health and Policy, London School of Hygiene and Tropical Medicine, London WC1E 7HT, UK
e-mail: alys.mcalpine@lshtm.ac.uk

J. Q. Smith
Department of Statistics, University of Warwick, Coventry CV4 7AL, UK

R. Argiento et al. (eds.), *New Frontiers in Bayesian Statistics*, Springer Proceedings in Mathematics & Statistics 405, https://doi.org/10.1007/978-3-031-16427-9_3

ing and exploitation. It is estimated that 23% of victims of forced labour [12] and 60% of victims of human trafficking were outside their country of residence [23]. In order to inform policymakers attempting to prevent exploitation, it is important to understand migrants' journeys and identify how individuals' hyper-precarity and livelihood insecurity, experienced due to both employment and immigration [14], evolves on different migration pathways.

Increasingly, Agent Based Models (ABMs) have been commonly used in contexts such as migration as they focus on the level of the individual and can be constructed by modelling the potential outcomes of successive events and decision-making [13, 15]. In order to construct these models, a range of data sources, such as large structured demographic datasets or natural language narratives and theories have been used to inform deterministic and stochastic transitions within an ABM. These transitions take the form of either mathematical equations, such as differential equations, or heuristic if-then rules and are informed by experts who describe the influences, possible options available and threats along their journey to another country. However there is often ambiguity in the reporting of these models. For this type of egocentric modelling with heterogeneous actors and actions, where the focus is on an individual's decisions, ABMs are an obvious choice and hence are being increasingly applied to model migration, though not yet with great detail on the true range of actors and decisions due to the complex nature of the application and difficulty in acquiring testimonies. Despite their increasing popularity, ABMs are unable to naturally combine expert judgement with available data to estimate and validate them. This is a problem as these steps are particularly important in this domain due to the previously mentioned difficulty in obtaining large amounts of data.

Chain Event Graphs (CEGs) are directed acyclic graphs that describe the evolution of a process through an unfolding of events [20]. CEGs are transformations on event trees and therefore are able to represent context-specific independence statements, conditional independence statements that are true only in specific contexts. The CEG should be thought of as a collection of *florets* (non-leaf nodes and their outgoing edges) that represent the events and their outcomes of the modelled process. The CEG represents the aforementioned independence statements by providing a staging on the florets that denotes their exchangeability. An example demonstrating these concepts is shown in Sects. 2.1 and 3.1. A particular class of CEGs, non-stratified CEGs, are able to more naturally represent an asymmetric unfolding of events. More generally, CEGs have previously been used for modelling in a wide range of applications, such as criminal collaborating [3], public health [18] and educational studies [6].

In this paper, we present a new methodology being developed to provide a Bayesian framework to an existing ABM through transforming it into a CEG. There are many key benefits of transforming the ABM into a CEG. One key advantage of the CEG is its compact representation, which not only shows the asymmetries in the events, as was the case with the initial diagram of the ABM, but also explicitly represents the context-specific conditional independences within the graph's topology. As a result, the potential series of events that may be experienced by a migrant and how these events impact future events are easily comprehensible. Secondly, by using the transformation of an ABM into a CEG we can apply a Bayesian framework in a natural way. This is particularly valuable in the situation where, due to the nature of migration data, the ability to perform Bayesian inference to combine data expressed through individual testimonies or expert descriptions is vital. Further benefits include the ability to use Bayesian model selection to compare the likelihood of different independence statements around the outcomes of events, represented by different theories of migration, using Bayes factor. For these reasons, the CEG makes a highly effective conduit into a stochastic description of the problem.

Standard structural models such as Bayesian Networks (BNs), a subset of CEGs [2], do not provide a good framework for egocentric modelling because the underlying processes and data tends to be highly asymmetrical and therefore does not allow a product space structure that is present in a BN. This is illustrated by the fact that ABMs—such as the ones used in the application above – typically need to use very different transitions depending on the current state the agent finds themselves in. BNs are also not able to represent context-specific independence statements where an independence relationship holds only for certain values of the conditioning variable. The presence of context-specific independence statements is also common in this application; examples of such statements are provided in our illustrative example.

This is the first paper that investigates how an ABM can be used to construct a CEG; it is the first genuine Bayesian model of migration processes to be built which draws from a combination of testimonies, surveys typical data and expert judgement. In Sect. 2, we give a background into ABMs of migration, formalise the class of models we are considering and introduce our illustrative example. In Sect. 3, we introduce the CEG, explain how it can represent the ABM and the benefits of this approach and continue our example of how to convert a given ABM into a CEG. We conclude, in Sect. 4, with a discussion of future work.

2 Agent Based Models of Migration

Migrants' pathways are often complex and non-linear, making many conventional modelling approaches unsuitable. ABMs provide a bottom-up approach to modelling, where the focus is on the individual. The aim of these models is to accurately replicate a population, its environment and the interactions that occur.

Despite their ability to plausibly model the transitions of an agent, many ABMs, both in migration research [15] and more broadly [7, 11], have been described as opaque with many of the critical details needed to fully understand or replicate the models missing from publication, such as the lack of standardising model development. Some attempts, such as the ODD protocol, have been made to create a standardised structure for explaining ABMs [7], but there is still significant variance in how the protocol is used and the clarity it brings to ABMs. ABMs' application often depends on the implementation of often severely constraining software which may or may not match the modelled domain well. Perhaps even more concerning is the gulf that exists when applying such models between the domain and a principled statistical inference about that domain. In particular no real guidance about how to set the ABM parameters is given, estimation of these is naive and model selection performed simply by matching trajectories of hypothesised models with chosen/estimated parameters with sampled trajectories. As a result, others [1, 8, 9, 16] have already identified the desperate need for embedding more principled ways of performing inference in order to estimate and validate ABM models when these are applied to real case studies. In this paper we argue that the best way of doing this is by using Bayesian models formulated around tree based CEG methods in ways we illustrate below.

As a first step we of course need to provide a proper formal systematic description of an ABM—something that is sadly missing from many applications of this promising technology. Here we follow [11] who express the ABM as a particular class of dynamic system model where agents are variables and their transitions are given by local updating functions. This work provides a similar statistical framework in order to study ABMs. We consider a set of agents $(x_1, x_2, \ldots, x_n)$ that take values in $\mathbb{S}$ a finite discrete set that represents the possible states that an agent can be in. The set of all possible values of all of the agents in the system gives the state-space. For any given state in the space, the updating process that determines the transitions between states is a Markov process. The possible transitions in the Markov process can be represented by a directed graph $G = (V, E)$ with V the state space and edges $e \in E$ between $u \in V$ and $v \in V$ if it is possible to transition from state u to v.

To provide a comprehensive translation of general ABMs as formally described above into Bayesian stochastic models would be a massive task and beyond the scope of this short paper. Here, for simplicity, we constrain our discussion to those ABMs with only one agent, and with a Markov process that has graph representation in the form of a finite, rooted, directed tree. The simplification of only using one agent is reasoned by the nature of these models being largely egocentric with the process and decision-making depending solely on the state of the individual, even if affected by interactions with other agents and the environment. The rationale of only allowing a finite, rooted, directed tree for the updating of states is justified: due to the nature of models of migrants pathways, we are interested in ABMs that can be thought of as an unfolding of a sequence of events. A tree gives the most natural representation of this process [17].

2.1 An Illustrative Example of Migrant Behaviour

Here we introduce an illustrative example ABM of an individual's decision on whether to migration or not represented in Fig. 1. This decision is modelled as a sequence of events that impact their final decision. In this example, the ABM starts by initialising an individual's socio-economic status, X_I. The individual then may receive an offer to migrate, X_O. This offer either comes with or without employment, X_E. Finally, the individual makes a decision as to whether they should migrate or not, X_M. Each of the nodes in this diagram has an if-then rule associated with its transitions. For instance, Fig. 1 shows an example heuristic rule for the decision

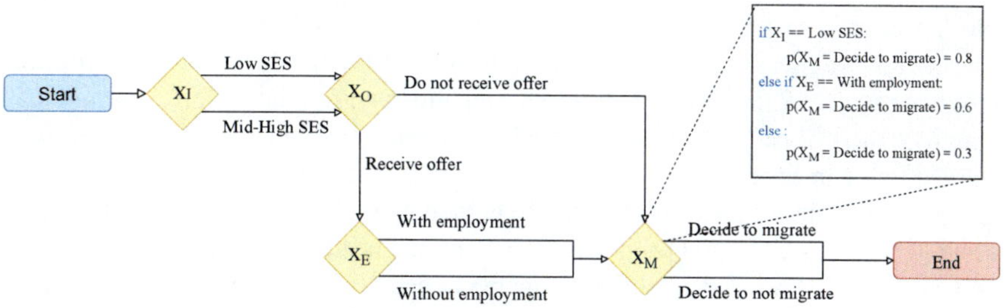

Fig. 1 Example of an agent based model for migration

to migrate. This rule shows how the probability of migrating is dependent on the outcomes of previous events.

3 From ABMs to CEGs

In Sect. 2, we described the class of ABMs that we are considering in this work. This decision is justified as the types of information we have about this process is best represented through a probability tree representing the possible progress of each migrant in the migrant population. This is particularly useful as it depicts the step-by-step nature of the process, where each migrant decides their next course of action. Typical hypotheses concerning this progress assume various conditional independence hypotheses, such as those shown in Sect. 3.1. Within an event tree model, these can be expressed by the stage structure on the florets of the tree.

The if-then rules within a heuristic, egocentric ABM implicitly include independence hypotheses regarding the outcome of an event for an individual through the choice of inputs considered. By assuming this conditional independence within a hypothesised model, we can identify those migrants within a sample who can be assumed on the next step of their journey to be exchangeable with each other. This is important if we wish to understand the processes of migration through the relationships between unfolding events, and crucial if we wish to understand the impact of potential targeted interventions. The CEG provides a framework in which to embed this model.

Bayesian methods are critical within such models because whenever models are sufficiently large to give a credible description of the processes, many parts of such processes are only sparsely observed. It is, therefore, critical to embed expert judgements through the priors on the hyperparameters. In this work, this is the distributions on the prior floret probabilities. In this way, our proposed methodology scales up to granularities of descriptions shared by the ABMs of such processes.

We can embed not only the prior expectations of these probabilities – as often needed in typical ABMs—but also their uncertainty. This embellishment means that, by using the exchangeability assumptions alluded to above and embedded in a

Bayesian model, we can perform a prior-to-posterior update on these probabilities. In particular, we can derive principled model selection algorithms that respect the relative security of knowledge of different transitions within the system, through the strength of the priors. We note that, even if no actual steps in some of the paths are observed, we can proceed with this inference, whilst if many people are observed making a particular collection of transitions then estimated transition probabilities will be close to their sample proportions. The model is suitably regularised.

Furthermore, if we assume floret independence, we can perform a conjugate Bayesian analysis (for full details, see [5] and [4]). The consequent Bayesian model estimation and selection is both transparent and rapid due to the closed form representation and the interpretative understanding of the hyperparameters.

In particular, assuming each transition is multinomially distributed over the set of outcomes, to perform a conjugate analysis, we need to set the Dirichlet priors. The distributions for the transition probabilities are often not elicited in advance, due to the non-Bayesian nature of ABMs. However, if the values elicited are the mean transition probabilities, we can use these values as the prior means for the Dirichlet prior. In order to get the full prior distribution, we must add in a count of effective sample size. This acts as a measure of strength of the beliefs held within the ABM. This can be done either by eliciting such a value or by completing a sensitivity analysis around the value chosen, similar to the method taken in [18]. Other methods for setting up the hyperparameters can be seen in [4].

In order to compare competing models we can set the hyperparameters so they match each other as closely as possible as in [10]. This is implemented via a mind experiment, where strengths of expert's elicited opinions are expressed using phantom samples over potential root to leaf path developments.

Of course, we could fit a CEG directly to model the migration process, through eliciting an event tree, the hypothesises and the prior distributions. However, if such an ABM has already been developed and thoughtfully calibrated to domain understanding—as is often the case—then it would be inefficient to ignore this information. As we can exploit the fact that the CEG is largely compatible with the ABM, it can be used to embellish the original, rather coarse, description given by the ABM into an inferential model which is fit for purpose.

3.1 Example Continued

Returning to our example, by untangling the current representation, we can obtain an event tree which is implied by the ABM. Within this class of ABM, an agent's transitions are determined by the outcomes of their previous transitions. Therefore, the next transition is conditional on its previous events. Such events define the situations (non-leaf nodes) in the CEG; there is a direct link between the CEG and the ABM. The nodes in the ABM define the situations in the CEG, with the possible transitions from that node represented by the floret around that situation. The event tree thus obtained is shown in Fig. 2. This is an example of an asymmetric unfolding

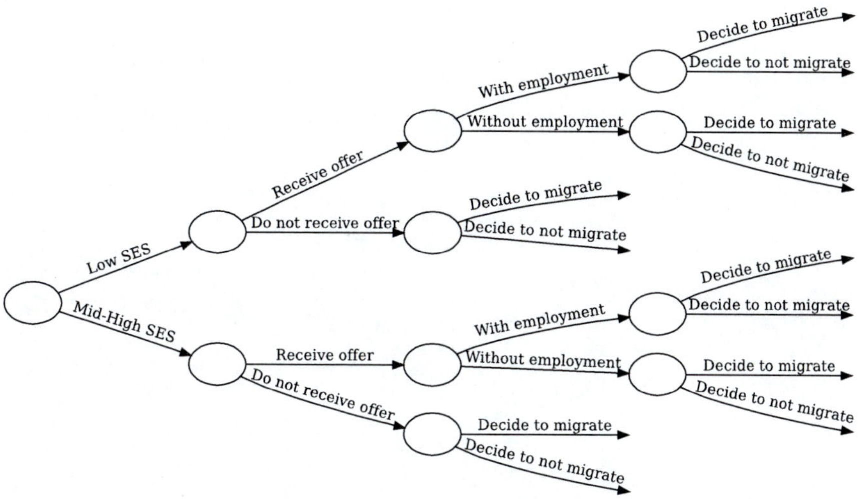

Fig. 2 Event tree representation of the ABM shown in Fig. 1. The leaf nodes are suppressed to prevent visual cluttering.

of events; if the migrant does not receive an offer to migrate, we do not need to consider whether the offer contains employment. This is denoted here as:

$$\nexists X_E | X_O = no. \tag{1}$$

Next, by looking at the if-then rules within the ABM, we can identify the implicit independence statements that exist within these rules. For the decision rule shown about the decision to migrate, we have the independence statements:

$$X_M \perp\!\!\!\perp X_O, X_E | X_I = low \tag{2}$$

$$X_M \perp\!\!\!\perp X_O | \{X_I = mid\text{-}high, X_E \neq yes\} \tag{3}$$

This provides the staging for the CEG. The staging can be represented by a staged tree, an event tree with florets in the same stage coloured the same. The staged tree for this example is shown in Fig. 3.

For this example, we assume that the other rules in the ABM represent the following statements:

- W_2 (Yellow): Regardless of socio-economic status, the probability of receiving an offer is the same.
- W_3 (Green): When an offer is received, the probability of it containing an employment contract is the same, irrespective of socio-economic status.

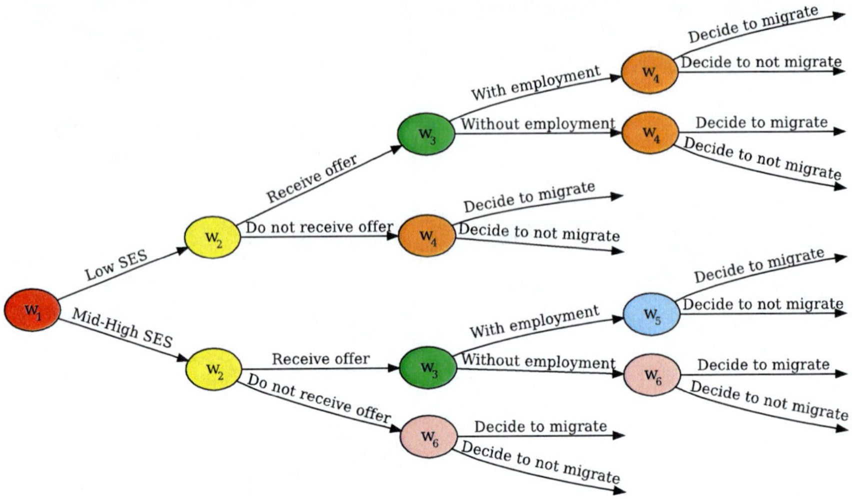

Fig. 3 Staged tree representation of the ABM. Here, 'SES' refers to socio-economic status. The leaf nodes are suppressed to prevent visual cluttering

- W_4 (Orange): A migrant with low socio-economic status has the same probability of deciding to migrate, irrespective of whether they have received an offer and whether their offer contained an employment contract.
- W_6 (Pink): A migrant with mid-high socio-economic status has the same probability of deciding to migrate if either (a) they receive an offer but it does not contain an employment contract or (b) they do not receive an offer in the first place.

From the staged tree, we can identify the nodes that are in the same position. In this example, w_4 and w_6 have the same future unfoldings for all future events, and are therefore in the same position.

Note that some nodes are the same stage but not the same position; w_3 is one such example, where the probability of the offer having employment is the same but the migrants' longer-term decision-making will still be influenced by their socio-economic status from earlier in the tree. This example demonstrates a context-specific independence statement: the decision to migrate is independent of whether you have an offer to migrate if your socio-economic status is low (Fig. 4).

This example shows the CEG can model and provide a compact representation of the conditional independence hypotheses present in the ABM. The transformation from the ABM into the CEG now enables the natural transformation of the model into a Bayesian framework with its associated previously described benefits.

4 Discussion

We have demonstrated that we are able to transform ABMs into CEGs. The benefits of this transformation are clear: it provides a compact representation of its independence statements, directly from the topology of the graph. This is valuable in identifying whether the model is making a plausible set of assumptions and making the independence structure accessible to be understood by those without a mathematical background, such as policymakers. The transformation into a CEG also allows for a natural conversion into a Bayesian framework with additional benefits: improved uncertainty quantification, Bayesian inference with available data and Bayesian model selection.

Whilst this paper specifically focuses on migration, CEGs have many potential applications in other domains where ABMs have been used to represent ego-based processes, such as dietary, voting or criminal behaviour.

This research reflects work in process; further investigation is needed to extend this methodology and increase the scope of ABMs that it applies to. Exploration of new representations is ongoing; one extension of the CEG could include the recently developed continuous time dynamic CEG, which is able to accommodate recurrent within the ABM structure and model holding times along the edges between events [19]. Further extensions of interest focus on CEGs which are able to represent the interactions of multi-agent systems players such as in [21], and agents looping through a CEG with changing probabilities over time depending on previous migration experience. Engaging with this research will provide many avenues of future research to build upon the work presented in this paper, enabling for more full and direct CEG-like representations of an even wider class of ABMs than those discussed above. The full results of this study will be published, alongside any future extensions, in a later paper.

Acknowledgements Peter Strong was supported by the EPSRC and the MRC [grant number EP/L015374/1]. Alys McAlpine was supported by UKRI [grant number ES/V006681/1] Jim Q. Smith was funded by the EPSRC [grant number EP/K03 9628/1]. We would like to thank Aditi Shenvi for her valuable comments.

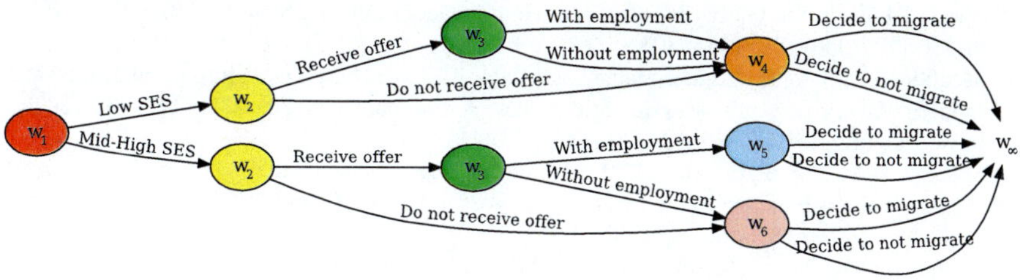

Fig. 4 A CEG representation of the above ABM with some examples of independence statements. 'SES' stands for socio-economic status

References

1. An, L., Grimm, V., Sullivan, A., Turner II, B., Malleson, N., Heppenstall, A., Vincenot, C., Robinson, D., Ye, X., Liu, J., et al.: Challenges, tasks, and opportunities in modeling agent-based complex systems. Ecol. Modell. (2021). https://www.sciencedirect.com/science/article/pii/S030438002100243X
2. Barclay, L., Hutton, J., Smith, J.Q.: Refining a Bayesian network using a chain event graph. Int. J. Approximate Reasoning **54**, 1300–1309 (2013). https://doi.org/10.1016/j.ijar.2013.05.006
3. Bunnin, F.O., Shenvi, A., Smith, J.Q.: Network modelling of criminal collaborations with dynamic Bayesian steady evolutions (2020). ArXiv preprint arXiv:2007.04410
4. Collazo, R.A., Görgen, C., Smith, J.Q.: Chain Event Graphs. CRC Press (2018)
5. Freeman, G., Smith, J.: Bayesian MAP model selection of chain event graphs. J. Multivar. Anal. **102**(7), 1152–1165 (2011). https://doi.org/10.1016/j.jmva.2011.03.008
6. Freeman, G., Smith, J.Q.: Dynamic staged trees for discrete multivariate time series: forecasting, model selection and causal analysis. Bayesian Anal. **6**(2) (2011). https://doi.org/10.1214/11-ba610
7. Grimm, V., Berger, U., Bastiansen, F., Eliassen, S., Ginot, V., Giske, J., Goss-Custard, J., Grand, T., Heinz, S.K., Huse, G., et al.: A standard protocol for describing individual-based and agent-based models. Ecol. Modell. **198**(1–2), 115–126 (2006). https://doi.org/10.1016/j.ecolmodel.2006.04.023
8. Grimm, V., Revilla, E., Berger, U., Jeltsch, F., Mooij, W.M., Railsback, S.F., Thulke, H.H., Weiner, J., Wiegand, T., DeAngelis, D.L., et al.: Pattern-oriented modeling of agent-based complex systems: lessons from ecology. Science (2005). https://www.science.org/doi/10.1126/science.1116681
9. Heckbert, S., Baynes, T., Reeson, A.: Agent-based modeling in ecological economics. Ann. N. Y. Acad. Sci. (2010). https://nyaspubs.onlinelibrary.wiley.com/doi/10.1111/j.1749-6632.2009.05286.x
10. Heckerman, D., Geiger, D., Chickering, D.M.: Learning Bayesian networks: the combination of knowledge and statistical data. Mach. Learn. **20**(3), 197–243 (1995). https://doi.org/10.1007/bf00994016
11. Hinkelmann, F., Murrugarra, D., Jarrah, A.S., Laubenbacher, R.: A mathematical framework for agent based models of complex biological networks. Bull. Math. Biol. **73**(7), 1583–1602 (2010). https://doi.org/10.1007/s11538-010-9582-8
12. International Labour Organisation: Global estimates of modern slavery: forced labour and forced marriage. Tech. Rep, International Labour Organisation (2017)
13. Klabunde, A., Willekens, F.: Decision-making in agent-based models of migration: state of the art and challenges. Eur. J. Popul. (2016). https://link.springer.com/article/10.1007/s10680-015-9362-0
14. Lewis, H., Peter, D., Hodkinson, S., Louise, W.: Hyper-precarious lives: Migrants, work and forced labour in the Global North. Prog. Human Geogr. **39**(5), 580–600 (2015). https://doi.org/10.1177/0309132514548303
15. Mcalpine, A., Kiss, L., Zimmerman, C., Chalabi, Z.: Agent-based modeling for migration and modern slavery research: a systematic review. J. Comput. Soc. Sci. **4**(1), 243–332 (2020). https://doi.org/10.1007/s42001-020-00076-7
16. Schulze, J., Müller, B., Groeneveld, J., Grimm, V.: Agent-based modelling of social-ecological systems: achievements, challenges, and a way forward. J. Artif. Soc. Soc. Simul. **20**(2) (2017). https://doi.org/10.18564/jasss.3423
17. Shafer, G.: The Art of Causal Conjecture. MIT Press (1996)
18. Shenvi, A., Smith, J.Q.: A Bayesian Dynamic Graphical Model for Recurrent Events in Public Health (2019). ArXiv preprint arXiv:1811.08872
19. Shenvi, A., Smith, J.Q.: Propagation for Dynamic Continuous Time Chain Event Graphs (2020). ArXiv preprint arXiv:2006.15865

20. Smith, J.Q., Anderson, P.E.: Conditional independence and chain event graphs. Artif. Intell. **172**(1), 42–68 (2008)
21. Thwaites, P.A., Smith, J.Q.: A graphical method for simplifying Bayesian games. Reliab. Eng. Syst. Saf. (2017). https://www.sciencedirect.com/science/article/pii/S0951832017305355
22. United Nations: The 17 goals | sustainable development. Tech. rep., U. N. (2021). https://sdgs.un.org/goals
23. United Nations Office on Drugs and Crime: global report on trafficking in persons 2016. Tech. rep., UNODC (2017). https://www.unodc.org/documents/data-and-analysis/glotip/2016_Global_Report_on_Trafficking_in_Persons.pdf

Power-Expected-Posterior Methodology with Baseline Shrinkage Priors

G. Tzoumerkas and D. Fouskakis

Abstract The Power-Expected-Posterior (PEP) prior gives us a convenient and objective method to deal with variable selection problems, under the Bayesian perspective, in regression models. The PEP prior inherits all of the advantages of Expected-Posterior-Prior (EPP) and furthermore it drops the need of selection over the imaginary data and decreases their effect over the final prior. Under the PEP prior methodology an initial (usually default) baseline prior is updated using imaginary data. This work focuses on normal regression models when the number of observations n is smaller than the number of explanatory variables p. We introduce the PEP prior methodology using different baseline shrinkage priors and we perform some comparisons in simulated data sets.

Keywords Bayesian variable selection · imaginary training sample · objective priors · shrinkage priors · sparse datasets

1 Introduction

We consider the variable selection problem for normal regression models, where the number of observations n is smaller than the number of explanatory variables p. Suppose the model space consists of all combinations of available covariates. Then for every model M_ℓ, in model space $\mathcal{M}$, the likelihood is given by

$$f_\ell(\boldsymbol{y}|X_\ell, \boldsymbol{\beta}_\ell, \sigma^2) = f_{N_n}(\boldsymbol{y}; X_\ell \boldsymbol{\beta}_\ell, \sigma^2 I_n),$$

where $f_{N_d}(\boldsymbol{y}; \boldsymbol{\mu}, \Sigma)$ is denoting the d-dimensional normal distribution with mean $\boldsymbol{\mu}$ and covariance matrix Σ. Furthermore, $\boldsymbol{y} = (y_1, \ldots, y_n)^T$ denotes the response data, X_ℓ is the $n \times p_\ell$ design matrix; where p_ℓ is the number of explanatory variables

G. Tzoumerkas (✉) · D. Fouskakis
Department of Mathematics, National Technical University of Athens, Athens, Greece
e-mail: tzoumg@mail.ntua.gr

D. Fouskakis
e-mail: fouskakis@math.ntua.gr

R. Argiento et al. (eds.), *New Frontiers in Bayesian Statistics*, Springer Proceedings in Mathematics & Statistics 405, https://doi.org/10.1007/978-3-031-16427-9_4

under model M_ℓ, $\boldsymbol{\beta}_\ell$ is a vector of length p_ℓ of the effects of each covariate on the response variable, I_n is the $n \times n$ identity matrix and σ^2 is the error variance. We assume that $\boldsymbol{y}$ and the columns of the design matrix of the full model (including all available explanatory variables) have been centered on zero, so there is no intercept in our model.

Under the Bayesian model choice perspective, we have to set priors both for the model space and the parameter space of each model. Regarding the prior on the model space, for sparsity reasons, we consider the uniform prior on model size, as a special case of the beta-binomial prior; see [18]. With respect to the prior distribution on the coefficients in each model, because we are not confident about any given set of regressors as explanatory variables, little prior information on their regression coefficients can be expected. This argument alone justifies the need for an objective model choice approach in which vague prior information is assumed. Furthermore, we need to use a prior capable to deal with the $n < p$ scenario. Finally, regarding the (common across models) error variance, the reference prior will be used, i.e. $\pi(\sigma^2) \propto \sigma^{-2}$.

1.1 Shrinkage Priors

A common way to deal with normal regression problems, when $n < p$, is by using shrinkage methods. Under the Bayesian perspective this can be done using a shrinkage prior on the model coefficients. By the term shrinkage, it is declared that the covariates that correspond to explanatory variables that do not affect the response variable will shrink towards zero. Shrinkage priors share eminent theoretical properties, compelling computational complexity and great empirical performance (e.g. [5, 17]).

A shrinkage prior can often be conceived as a scale-mixture prior, which is placed on the regression coefficients of every possible model. Something that characterizes such shrinkage priors, is their hyperparameters: the global shrinkage hyperparameter, that determines the overall sparsity in the whole parameter vector and the local shrinkage hyperparameter, where a distinct shrinkage parameter is considered specifically for every single effect and controls the shrinkage of this individual effect. Depending on the shrinkage prior, the global parameter or the local parameters may be absent from the formation.

By assuming a shrinkage prior, on the vector of regression coefficients $\boldsymbol{\beta}_\ell$, in most of the cases a prior with heavy mass around zero is being produced and by so, non-true effects shrink towards zero. Furthermore, heavy tails are important, as they avert true effects to get shrinked. In Table 1, we mention some, often used, shrinkage priors, where by τ we refer to local shrinkage hyperparameters and by λ to global shrinkage hyperparameters. In all of the cases that a global shrinkage hyperparameter exists in the formation of a shrinkage prior (except Ridge g-prior), we consider a half-Cauchy prior on λ, which is a common choice in Bayesian hierarchical models (e.g. [17]). Furthermore, except Ridge g-prior, independent conditional priors for the

Table 1 A list of shrinkage priors

#	Name	Conditional prior of $\boldsymbol{\beta}_\ell$	Shrinkage hyparameters
1	LASSO [15]	$\beta_j \mid \tau_j^2, \sigma^2 \sim N(0, \sigma^2\tau_j^2)$	$\tau_j^2 \mid \lambda \sim Exp(\frac{\lambda^2}{2})$ $\lambda \sim HC(0, 1)^a$
2	Horseshoe [2]	$\beta_j \mid \lambda, \tau_j, \sigma^2 \sim N(0, \sigma^2\lambda^2\tau_j^2)$	$\tau_j \sim HC(0, 1)$ $\lambda \sim HC(0, 1)$
3	Ridge [11]	$\beta_j \mid \lambda, \sigma^2 \sim N(0, \sigma^2\frac{1}{\lambda})$	$\lambda \sim HC(0, 1)$
4	Local Student's t [20]	$\beta_j \mid \tau_j^2, \sigma^2 \sim N(0, \sigma^2\tau_j^2)$	$\tau_j^2 \mid \lambda \sim IG(\frac{k}{2}, \frac{k}{2\lambda})^b$ $\lambda \sim HC(0, 1)$ k fixed
5	Elastic Net [13]	$\beta_j \mid \lambda_2, \tau_j, \sigma^2 \sim N(0, \sigma^2\frac{1}{\lambda_2+\tau_j^2})$	$\tau_j^2 \mid \lambda_1 \sim Exp(\frac{\lambda_1^2}{2})$ $\lambda_1, \lambda_2 \sim HC(0, 1)$
6	Beta Prime [1]	$\beta_j \mid \tau_j^2, \sigma^2 \sim N(0, \tau_j^2\sigma^2)$	$\tau_j^2 \sim Inv-Beta(a, b)$ a, b fixed
7	Ridge g-prior [10]	$\boldsymbol{\beta}_\ell \mid \lambda, \sigma^2 \sim N_{p_\ell}(\mathbf{0}, \sigma^2 V_\ell)$, $V_\ell = g(X_\ell^T X_\ell + \lambda I_{p_\ell})^{-1}$	$g = max\{n, p_\ell^2\}$, λ fixed

[a] $HC(x_0, \gamma)$, (half-Cauchy) is the truncated Cauchy distribution with location parameter x_0, scale parameter γ and support (x_0, ∞)
[b] $IG(\alpha, \beta)$, denotes the Inverse Gamma distribution, with shape parameter α and scale parameter β

coefficients of model M_ℓ are used and therefore, for those cases, we only present the marginal prior for $j = 1, \ldots, p_\ell$.

1.2 *Power-Expected-Posterior Priors*

A principal approach to define objective priors is the use of random imaginary training data [4]. Power-Expected-Posterior (PEP) prior [6, 7], uses this methodology. In particular the PEP prior is defined as

$$\pi_\ell^{PEP}(\boldsymbol{\beta}_\ell \mid \sigma^2, \delta, X_\ell^*) = \int \pi_\ell^N(\boldsymbol{\beta}_\ell \mid \mathbf{y}^*, \sigma^2, \delta, X_\ell^*) m_0^N(\mathbf{y}^* \mid \sigma^2, \delta, X_0^*) d\mathbf{y}^*, \quad (1)$$

$$\pi_\ell^{PEP}(\sigma^2) = \pi^N(\sigma^2) \propto \frac{1}{\sigma^2},$$

with

$$\pi_\ell^N(\boldsymbol{\beta}_\ell \mid \mathbf{y}^*, \sigma^2, \delta, X_\ell^*) \propto f_\ell(\mathbf{y}^* \mid \boldsymbol{\beta}_\ell, \sigma^2, \delta, X_\ell^*) \pi_\ell^N(\boldsymbol{\beta}_\ell \mid \sigma^2, X_\ell^*) \quad (2)$$

and

$$f_\ell(\boldsymbol{y}^*|\boldsymbol{\beta}_\ell, \sigma^2, \delta, X_\ell^*) = \frac{f_\ell(\boldsymbol{y}^*|\boldsymbol{\beta}_\ell, \sigma^2, X_\ell^*)^{1/\delta}}{\int f_\ell(\boldsymbol{y}^*|\boldsymbol{\beta}_\ell, \sigma^2, X_\ell^*)^{1/\delta} d\boldsymbol{y}^*}. \quad (3)$$

In the above equations, we have set $\boldsymbol{y}^*$ to be the imaginary observations of size n^* and X_ℓ^* the imaginary design matrix of model M_ℓ. By $\pi_\ell^N(\boldsymbol{\beta}_\ell|\boldsymbol{y}^*, \sigma^2, \delta, X_\ell^*)$ we denote the conditional on σ^2 posterior of $\boldsymbol{\beta}_\ell$, using a baseline prior $\pi_\ell^N(\boldsymbol{\beta}_\ell|\sigma^2, X_\ell^*)$ and data $\boldsymbol{y}^*$. In equation (3) the likelihood of imaginary observations is raised to the power of $1/\delta$ and density normalized. By doing this we decrease the effect of the imaginary data. For $\delta = 1$, Eq. (1) results to the Expected-Posterior-Prior (EPP) [16]. In order to have a unit information interpretation [12], we could set $\delta = n^*$ and in order to avoid any effect of the choice of imaginary design matrices, we set $n^* = n$ and we have that $X_\ell^* = X_\ell$. In Eq. (1), $m_0^N(\boldsymbol{y}^*|\sigma^2, \delta, X_0^*)$, is the prior predictive distribution (or the marginal likelihood), evaluated at $\boldsymbol{y}^*$, of the reference model M_0, given σ^2. As a reference model we consider, for reasons of parsimony, the model with no covariates (null model). Finally, for every model M_ℓ, the marginal likelihood under the baseline prior is given by

$$m_\ell^N(\boldsymbol{y}^*|\sigma^2, \delta, X_\ell^*) = \int f_\ell(\boldsymbol{y}^*|\boldsymbol{\beta}_\ell, \sigma^2, \delta, X_\ell^*)\pi_\ell^N(\boldsymbol{\beta}_\ell|\sigma^2, X_\ell^*)d\boldsymbol{\beta}_\ell. \quad (4)$$

2 PEP-Shrinkage Prior

In the above formulation, by choosing as a baseline prior $\pi_\ell^N(\boldsymbol{\beta}_\ell|\sigma^2, X_\ell^*)$ a shrinkage prior (see Table 1), a PEP-Shrinkage prior is created and thus we can apply the PEP prior methodology in shrinkage problems.

PEP priors can be considered as fully automatic, objective Bayesian methods for model comparison in regression models (see for example [4, 6]). They are developed through the utilization of the device of "imaginary" samples, coming from the simplest model under comparison. Therefore, PEP priors offer several advantages, among which they have an appealing interpretation based on imaginary training data coming from a prior predictive distribution and also provide an effective way to establish compatibility of priors among models (see [3]), through their dependence on a common marginal data distribution. Thus, the PEP methodology can be applied also with proper baseline prior distributions. Furthermore, by choosing the simplest model, as a reference model, to generate the imaginary samples, the PEP prior shares common ideas with the skeptical-prior approach described by Spiegelhalter et al. [19].

Under Eq. (3) the likelihood of the imaginary data $\boldsymbol{y}^*$, under model M_ℓ, is given by

$$f_\ell(\boldsymbol{y}^*|X_\ell^*, \boldsymbol{\beta}_\ell, \sigma^2, \delta) = f_{N_{n^*}}(\boldsymbol{y}^*; X_\ell^* \boldsymbol{\beta}_\ell, \delta\, \sigma^2\, I_{n^*}).$$

From Table 1 it is obvious that all shrinkage priors that we will use as baseline priors under the PEP methodology, have the following general form

$$\pi_\ell^N(\boldsymbol{\beta}_\ell|\boldsymbol{\theta}_\ell, \sigma^2) = f_{N_{p_\ell}}(\boldsymbol{\beta}_\ell; \mathbf{0}, \sigma^2\Omega_\ell),$$

where $\Omega_\ell \equiv \Omega_\ell(\boldsymbol{\theta}_\ell)$ is a $p_\ell \times p_\ell$ matrix, where its i-th main diagonal element is written as an equation of the global and the i-th local shrinkage hyperparameters. By $\boldsymbol{\theta}_\ell$ we denote the vector containing all the shrinkage hyperparameters of model M_ℓ, with a prior distribution denoted by $\pi(\boldsymbol{\theta}_\ell)$.

2.1 Conditional PEP-Shrinkage Prior

The conditional posterior distribution $\pi_\ell^N(\boldsymbol{\beta}_\ell|\boldsymbol{y}^*, \sigma^2, \delta, X_\ell^*, \boldsymbol{\theta}_\ell)$, using the baseline prior and the imaginary data is given by

$$\begin{aligned}\pi_\ell^N(\boldsymbol{\beta}_\ell|\boldsymbol{y}^*, \sigma^2, \delta, X_\ell^*, \boldsymbol{\theta}_\ell) &\propto f_\ell(\boldsymbol{y}^*|X_\ell^*, \boldsymbol{\beta}_\ell, \sigma^2, \delta)\pi_\ell^N(\boldsymbol{\beta}_\ell|\boldsymbol{\theta}_\ell, \sigma^2)\\ &= f_{N_{n^*}}(\boldsymbol{y}^*; X_\ell^* \,\boldsymbol{\beta}_\ell, \delta\, \sigma^2\, I_{n^*}) f_{N_{p_\ell}}(\boldsymbol{\beta}_\ell; \mathbf{0}, \sigma^2\Omega_\ell)\end{aligned}$$

and so we have have that

$$\pi_\ell^N(\boldsymbol{\beta}_\ell|\boldsymbol{y}^*, \sigma^2, \delta, X_\ell^*, \boldsymbol{\theta}_\ell) = f_{N_{p_\ell}}(\boldsymbol{\beta}_\ell; \delta^{-1}W_\ell X_\ell^{*T}\boldsymbol{y}^*, \sigma^2 W_\ell),$$

where $W_\ell = [\delta^{-1}X_\ell^{*T}X_\ell^* + \Omega_\ell^{-1}]^{-1}$. Moreover, from Eq. (4), for any model M_ℓ, the prior predictive distribution, under the baseline prior, conditional on σ^2 and $\boldsymbol{\theta}_\ell$ is

$$m_\ell^N(\boldsymbol{y}^*|\sigma^2, \delta, X_\ell^*, \boldsymbol{\theta}_\ell) = f_{N_{n^*}}(\boldsymbol{y}^*; \mathbf{0}, \sigma^2\Lambda_\ell),$$

where $\Lambda_\ell = X_\ell^*\Omega_\ell X_\ell^{*T} + \delta I_n$. Thus, the conditional PEP-Shrinkage prior is

$$\begin{aligned}\pi_\ell^{PEP}(\boldsymbol{\beta}_\ell|\sigma^2, \delta, X_\ell^*, \boldsymbol{\theta}_\ell) &= \int \pi_\ell^N(\boldsymbol{\beta}_\ell|\boldsymbol{y}^*, \sigma^2, \delta, X_\ell^*, \boldsymbol{\theta}_\ell) m_0^N(\boldsymbol{y}^*|\sigma^2, \delta, X_0^*) d\boldsymbol{y}^*\\ &= \int f_{N_{p_\ell}}(\boldsymbol{\beta}_\ell; \delta^{-1}W_\ell X_\ell^{*T}\boldsymbol{y}^*, \sigma^2 W_\ell) f_{N_n}(\boldsymbol{y}^*; \mathbf{0}, \sigma^2\Lambda_0) d\boldsymbol{y}^*\end{aligned}$$

and therefore we have that

$$\pi_\ell^{PEP}(\boldsymbol{\beta}_\ell|\sigma^2, \delta, X_\ell^*, \boldsymbol{\theta}_\ell) = f_{N_{p_\ell}}(\boldsymbol{\beta}_\ell; \mathbf{0}, \sigma^2 V_\ell),$$

where $V_\ell = [W_\ell^{-1} - \delta^{-2}X_\ell^{*T}Z_\ell X_\ell^*]^{-1}$ and $Z_\ell = [\delta^{-2}X_\ell^* W_\ell X_\ell^{*T} + \Lambda_0^{-1}]^{-1}$.

2.2 Conditional Posterior Under the PEP-Shrinkage Prior

The posterior distribution, under the PEP prior, conditional on the shrinkage hyper-parameters $\boldsymbol{\theta}_\ell$ of model M_ℓ, is given by

$$\begin{aligned}\pi_\ell^{PEP}(\boldsymbol{\beta}_\ell, \sigma^2|\mathbf{y}, \delta, X_\ell^*, X_\ell, \boldsymbol{\theta}_\ell) &\propto \pi_\ell^{PEP}(\boldsymbol{\beta}_\ell|\sigma^2, \delta, X_\ell^*, \boldsymbol{\theta}_\ell)\pi^N(\sigma^2) f_\ell(\mathbf{y}|X_\ell, \boldsymbol{\beta}_\ell, \sigma^2)\\ &= f_{N_{p_\ell}}(\boldsymbol{\beta}_\ell; \mathbf{0}, \sigma^2 V_\ell)\pi^N(\sigma^2) f_{N_n}(\mathbf{y}; X_\ell\, \boldsymbol{\beta}_\ell, \sigma^2\, I_n).\end{aligned}$$

Using the reference prior for σ^2 (see Sect. 1), this joint posterior can be written as the product of

$$\pi_\ell^{PEP}(\boldsymbol{\beta}_\ell|\mathbf{y}, \sigma^2, \delta, X_\ell^*, X_\ell, \boldsymbol{\theta}_\ell) = f_{N_{p_\ell}}(\boldsymbol{\beta}_\ell; S_\ell\, X_\ell^T\mathbf{y}, \sigma^2 S_\ell)$$

and

$$\pi_\ell^{PEP}(\sigma^2|\mathbf{y}, \delta, X_\ell^*, X_\ell, \boldsymbol{\theta}_\ell) = f_{IG}(\sigma^2; \alpha_\ell, b_\ell),$$

where $f_{IG}(x; \alpha, b)$ is denoting the Inverse Gamma distribution with shape parameter α and scale parameter b. Furthermore, we have set $S_\ell = (V_\ell^{-1} + {X_\ell}^T X_\ell)^{-1}, \alpha_\ell = \frac{n}{2}$ and $b_\ell = \frac{\mathbf{y}^T[I_n + X_\ell V_\ell {X_\ell}^T]^{-1}\mathbf{y}}{2}$.

2.3 Marginal Likelihood Under the PEP-Shrinkage Prior

The marginal likelihood, of model M_ℓ, under the PEP-Shrinkage prior, given the shrinkage parameter $\boldsymbol{\theta}_\ell$ is given by

$$\begin{aligned}m_\ell^{PEP}(\mathbf{y}|\delta, X_\ell^*, X_\ell, \boldsymbol{\theta}_\ell) &= \int \pi_\ell^{PEP}(\boldsymbol{\beta}_\ell|\sigma^2, \delta, X_\ell^*, \boldsymbol{\theta}_\ell)\pi^N(\sigma^2) f_\ell(\mathbf{y}|X_\ell, \boldsymbol{\beta}_\ell, \sigma^2) d\boldsymbol{\beta}_\ell\, d\sigma^2\\ &\propto (det(I_n + X_\ell V_\ell {X_\ell}^T))^{-\frac{1}{2}}(\mathbf{y}^T[I_n + X_\ell V_\ell {X_\ell}^T]^{-1}\mathbf{y})^{-\frac{n}{2}}.\end{aligned}$$

Therefore in cases where the shrinkage parameters of the baseline prior are fixed (e.g. Ridge g-prior), the above marginal likelihood can be calculated in closed form. The unknown normalizing constant, in the above expression, comes from the improper prior of the error variance, which is common in all compared models, and therefore we do not face any indeterminacy issues when calculating the Bayes factor.

When the shrinkage parameters are not fixed, the marginal likelihood is given by

$$m_\ell^{PEP}(\mathbf{y}) \equiv m_\ell^{PEP}(\mathbf{y}|\delta, X_\ell^*, X_\ell) = \int m_\ell^{PEP}(\mathbf{y}|\delta, X_\ell^*, X_\ell, \boldsymbol{\theta}_\ell)\pi(\boldsymbol{\theta}_\ell) d\boldsymbol{\theta}_\ell.$$

If the dimension of $\boldsymbol{\theta}_\ell$ is one (e.g. Ridge prior) the above integral can be easily numerically evaluated. Furthermore, in order to search the model space, MC^3 procedures [14] can be performed. If the dimension of $\boldsymbol{\theta}_\ell$ is greater than one (e.g. Horseshoe prior), we perform an MC^3 procedure, conditionally on $\boldsymbol{\theta}_\ell$, as in Algorithm 3 of the Appendix of [9], where each component of $\boldsymbol{\theta}_\ell$ is generated from its full conditional posterior distribution using a Metropolis-Hastings step.

3 Simulation Study

In this section we test the PEP-Shrinkage methodology (with $\delta = n = n^*$, $X_\ell^* = X_\ell$ and the reference model to be the null one) on simulated data. We use as a baseline prior, all the shrinkage priors listed in Table 1 and compare their results. Moreover we compare the results under the PEP-Ridge prior with the ones obtain by using the Ridge prior, without the PEP methodology.

We have simulated 100 different samples of length $n = 25$ with $p = 50$ predictors. The values of the explanatory variables have been generated from $N_{50}(\mathbf{0}, \Sigma)$, where the symmetrical matrix Σ has elements $\Sigma_{i,j} = (0.75)^{|i-j|}, i, j = 1, \ldots, 50$. Finally, we center the columns of the design matrix on zero. For the predictor effects we have set $(\beta_1, \beta_2, \beta_{10})^T = (2, 0.8, 1.5)^T$ and for all of the rest, we set to be equal to 0. We have set $\boldsymbol{y} = X\boldsymbol{\beta} + \boldsymbol{\epsilon}$, where $\boldsymbol{\epsilon} \sim N_{25}(\mathbf{0}, \sigma^2 I_{25})$, for $\sigma^2 = 1.5$. Finally, we center the values of the response variable on zero.

In Fig. 1 (left), we present the boxplots of the marginal posterior inclusion probabilities, for the true effects, of the 100 different samples, for the seven different PEP-Shrinkage priors. Regarding the two most influential variables, X_1 and X_{10}, under every baseline prior, we obtained high posterior inclusion probabilities with the majority of cases to be above 0.5. Furthermore, for these two effects, PEP-Ridge seems to outperform every other PEP-Shrinkage prior. On the contrary, PEP-(Ridge) g-prior seems to give the least satisfactory results. For the predictor X_2, the median marginal posterior inclusion probabilities are above 0.5, for all baseline priors, except one. As before, PEP-Ridge gives the most satisfactory results, while PEP-(Ridge) g-prior produces posterior inclusion probabilities with a median value below 0.5. For the non-true effects, for brevity reasons, we present results in Fig. 1 (right) only for a subset of them. More specifically we present results only for variables X_3, X_9 and X_{11}, which are the ones with the higher correlations with the true effects. For every selection of baseline prior, the median marginal posterior inclusion probabilities are below 0.5. It is distinct that, regardless the baseline prior we choose, only in a small percentage of occasions, the non-true effects would have been accepted as true effects of the model (posterior inclusion probabilities above 0.5). We notice that PEP-Ridge manages to give, in general, very small posterior inclusion probabilities with small variability also. For the rest of the non-true effects we get similar results.

In Fig. 2, we present the boxplots of the posterior inclusion probabilities of the true main effects (left) and the (previously made) selection of non-true effects (right) between the PEP-Ridge and the Ridge prior (without applying the PEP methodology). As for the true effects we notice similar results, as both priors manages to accept the true effects, in the vast majority of the cases. For predictor X_{10} we can observe slightly better results under the PEP-Ridge methodology. As for the non-true effects,

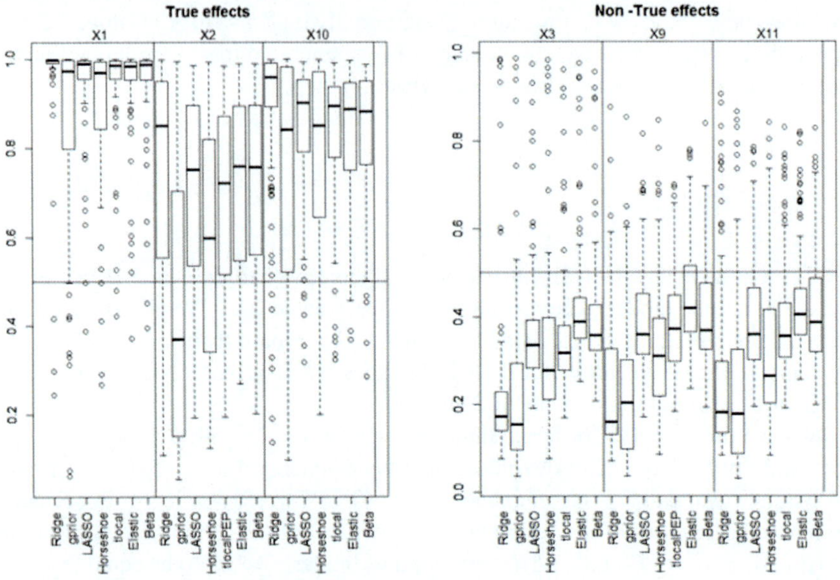

Fig. 1 Boxplots of posterior inclusion probabilities, across 100 simulated datasets, for the true effects—variables X_1, X_2, X_{10} (left) and for some of the non-true effects—variables X_3, X_9, X_{11} (right) using the PEP-Shrinkage methodology, for different baseline prior (X-axis).

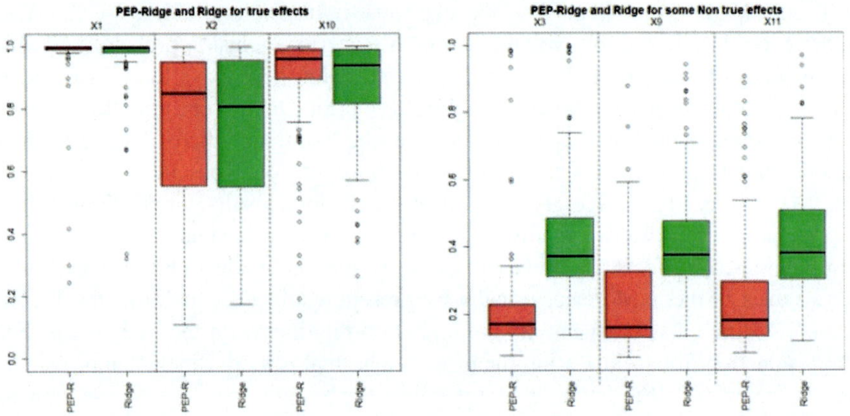

Fig. 2 Boxplots of posterior inclusion probabilities, across 100 simulated datasets, for the true effects—variables X_1, X_2, X_{10} (left) and for some of the non-true effects—variables X_3, X_9, X_{11} (right), using the PEP-Ridge prior (PEP-R) and the Ridge prior without the PEP methodology

the PEP-Ridge prior outperforms the Ridge prior, as it manages to restrains more cases to the desirable limits, that is, producing marginal posterior probabilities far below 0.5 with small variability. Thus we can conclude that the PEP methodology improves the initially chosen Ridge prior, as it produces more parsimonious results.

4 Discussion

In this paper we briefly present the model formulation and some preliminary results of an objective Bayesian prior distribution capable of dealing with variable selection problems in normal regression models when the number of observations is smaller than the number of explanatory variables. The proposed PEP-Shrinkage prior combines two approaches: the PEP prior methodology and the shrinkage priors. The resulting prior has a nice interpretation, based on imaginary data, and is compatible across models. Based on the simulation study, presented here, the PEP-Shrinkage priors, in the majority of cases, correctly identify the true model. Furthermore, under the Ridge prior, the PEP methodology improves the initial prior, by being more parsimonious, a property that is desirable on sparse regression problems.

There are several directions of future extensions. The main aim is to create a unified approach; i.e. a new class of PEP-Shrinkage priors, that includes all the cases mentioned in this paper. To achieve this goal our aim is to write the PEP-Shrinkage prior as a scale mixture of normal distribution, with the mixing distribution denoting the different baseline prior distributions used. This representation will offer several advantages: faster evaluation of posterior distributions and Bayes factors, under all approaches considered, as well as, computational tractability. The performance of this new class of shrinkage prior distributions then have to be assessed in relation to: (a) computational efficiency, (b) frequentist assessment, especially in terms of the speed of concentration of the posterior parameter distribution, or functional thereof, to the true value, and in terms of coverage of credible sets, (c) ease of interpretation, (d) default set of tuning hyperparameters in scientific applications. Moreover, a very important aspect is to check and prove mathematical properties of the new class of prior distributions. Further research should be held, of what happens if we choose the size of the imaginary data, not to be equal to the number of the observations and how that affects the results. In the same manner, we should check what happens for different values of δ, or even set a prior distribution for it, as in [8]. Finally, more shrinkage methods could be considered, apart the ones presented in Table 1. Additional future extensions of our PEP-Shrinkage method include implementation in generalized linear models, where computation is more demanding.

Acknowledgements This work has received funding from the Research Program PEVE 2020 of the National Technical University of Athens.

References

1. Bai, R., Ghosh, M.: On the beta prime prior for scale parameters in high-dimensional bayesian regression models. Stat. Sin. **31**, 843–865 (2021)
2. Carvalho, C.M., Polson, N.G., Scott, J.G.: The horseshoe estimator for sparse signals. Biometrika. **97**, 465–480 (2010)
3. Consonni, G., Veronese, P.: Compatibility of prior specifications across linear models. Stat. Sci. **23**, 332–353 (2008)
4. Consonni, G., Fouskakis, D., Liseo, B., Ntzoufras, I.: Prior Distributions for objective Bayesian analysis. Bayesian Anal. **13**, 627–679 (2018)
5. Datta, J., Ghosh, J.K.: Asymptotic properties of Bayes risk for the horseshoe prior. Bayesian Anal. **8**, 111–132 (2013)
6. Fouskakis, D., Ntzoufras, I., Draper, D.: Power-expected-posterior priors for variable selection in Gaussian linear models. Bayesian Anal. **10**, 75–107 (2015)
7. Fouskakis, D., Ntzoufras, I.: Power-conditional-expected priors. Using g-priors with random imaginary data for variable selection. J. Comput. Graph. Stat. **25**, 647–664 (2016)
8. Fouskakis, D., Ntzoufras, I., Perrakis, K.: Power-expected-posterior priors in generalized linear models. Bayesian Anal. **13**, 721–748 (2018)
9. Fouskakis, D., Ntzoufras, I.: Power-expected-posterior priors as mixtures of g-Priors. Bayesian Anal. (accepted) (2021)
10. Gupta, M., Ibrahim, J.: An information matrix prior for Bayesian analysis in generalized linear models with high dimensional data. Stat. Sin. **19**, 1641–1663 (2009)
11. Hsiang, T.C.: A Bayesian view on ridge regression. The Statist. **24**, 267–268 (1975)
12. Kass, R.E., Wasserman, L.: A reference Bayesian test for nested hypotheses and its relationship to the Schwarz criterion. J. Am. Stat. Assoc. **90**, 928–934 (1995)
13. Kyung, M., Gill, J., Ghosh, M., Casella, G.: Penalized regression, standard errors, and Bayesian lassos. Bayesian Anal. **5**, 369–411 (2010)
14. Madigan, D., York, J.: Bayesian graphical models for discrete data. Int. Stat. Rev. **63**, 215–232 (1995)
15. Park, T., Casella, G.: The Bayesian lasso. J. Am. Stat. Assoc. **103**, 681–687 (2008)
16. Pèrez, J.M., Berger, J.O.: Expected—posterior prior distributions for model selection. Biometrika **89**, 491–511 (2002)
17. Polson, G., Scott, J.: On the half-Cauchy prior for a global scale parameter. Bayesian Anal. **7**, 887–902 (2011)
18. Scott, J.G., Berger, J.O.: Bayes and empirical-Bayes multiplicity adjustment in the variable-selection problem. Ann. Stat. **38**, 2587–2619 (2010)
19. Spiegelhalter, D.J., Abrams, K.R., Myles, J.P.: Bayesian Approaches to Clinical Trials and Health-Care Evaluation. Wiley, Chichester (2004)
20. Tipping, M.E.: Sparse Bayesian Learning and the Relevance Vector Machine. J. Mach. Learn. **1**, 211–244 (2001)

Bayesian Nonparametric Scalar-on-Image Regression via Potts-Gibbs Random Partition Models

Mica Shu Xian Teo and Sara Wade

Abstract Scalar-on-image regression aims to investigate changes in a scalar response of interest based on high-dimensional imaging data. We propose a novel Bayesian nonparametric scalar-on-image regression model that utilises the spatial coordinates of the voxels to group voxels with similar effects on the response to have a common coefficient. We employ the Potts-Gibbs random partition model as the prior for the random partition in which the partition process is spatially dependent, thereby encouraging groups representing spatially contiguous regions. In addition, Bayesian shrinkage priors are utilised to identify the covariates and regions that are most relevant for the prediction. The proposed model is illustrated using the simulated data sets.

Keywords Bayesian nonparametric · Gibbs-type priors · Potts model · Clustering · Generalised Swendsen-Wang · High-dimensional imaging data

1 Introduction

Through advances in data acquisition, vast amounts of high-dimensional imaging data are collected to study phenomena in many fields. Such data are common in biomedical studies to understand a disease or condition of interest [2, 5, 39, 44], and in other fields such as psychology [3, 42], social sciences [7, 15, 17, 38], economics [12, 26, 27], climate sciences [30, 31], environmental sciences [4, 11, 22] and more. While extracting features from the images based on predefined regions of interest favours interpretation and eases computational and statistical issues, changes may occur in only part of a region or span multiple structures. In order to capture the

M. S. X. Teo · S. Wade (✉)
School of Mathematics and Maxwell Institute for Mathematical Sciences, University of Edinburgh James Clerk Maxwell Building, Edinburgh, UK
e-mail: sara.wade@ed.ac.uk

M. S. X. Teo
e-mail: mica.teo@ed.ac.uk

R. Argiento et al. (eds.), *New Frontiers in Bayesian Statistics*, Springer Proceedings in Mathematics & Statistics 405, https://doi.org/10.1007/978-3-031-16427-9_5

complex spatial pattern of changes and improve accuracy and understanding of the underlying phenomenon, sophisticated approaches are required that utilize the entire high-dimensional imaging data. However, the massive dimension of the images, which is often in the millions, combined with the relatively small sample size, which at best is usually in the hundreds, pose serious challenges.

In the statistical literature, this is framed as a scalar-on-image regression (SIR) problem [10, 14, 16, 19]. SIR belongs to the "large p, small n" paradigm; thus, many SIR models utilise shrinkage methods that additionally incorporate the spatial information in the image [10, 14, 16, 18, 19, 24, 37, 40, 46]. In the SIR problem, the covariates represent the image value at a single pixel/voxel, i.e. a very tiny region, and the effect on the response is most often weak, unreliable and difficult to interpret. Moreover, neighbouring pixels/voxels are highly correlated, making standard regression methods, even with shrinkage, problematic due to multicollinearity.

To overcome these difficulties, we develop a novel Bayesian nonparametric (BNP) SIR model that extracts interpretable and reliable features from the images by grouping voxels with similar effects on the response to have a common coefficient. Specifically, we employ the Potts-Gibbs model [21] as the prior of the random image partition to encourage spatially dependent clustering. In this case, features represent regions that are automatically defined to be the most discriminative. This not only improves the signal and eases interpretability, but also reduces the computational burden by drastically decreasing the image dimension and addressing the multicollinearity problem. Moreover, it allows sharp discontinuities in the coefficient image across regions, which may be relevant in medical applications to capture irregularities [46].

In this direction, [19] proposed the Ising-DP SIR model, which combines an Ising prior to incorporate the spatial information in the sparsity structure with a Dirichlet Process (DP) prior to group coefficients. Still, the spatial information is only incorporated in the sparsity structure and not in the BNP clustering model, which could result in regions that are dispersed throughout the image. Instead, we propose to incorporate the spatial information in the random partition model, encouraging spatially contiguous regions. Further advantages of the nonparametric model include a data-driven number of clusters, interpretable parameters, and efficient computations. Moreover, we combine this with heavy-tailed shrinkage priors [41] to identify relevant covariates and regions.

The remainder of this article is organized as follows. Section 2 outlines the development of the SIR model based on the Potts-Gibbs models. Section 3 derives the MCMC algorithm for posterior inference using the generalized Swendsen-Wang (GSW) [47] algorithm for efficient split-merge moves that take advantage of the spatial structure. Section 4 illustrates the methods through simulation studies. Section 5 concludes with a summary and future work.

2 Model Specification

We introduce the statistical models that form the basis of the proposed Potts-Gibbs SIR model: SIR, random image partition model and shrinkage prior.

2.1 *Scalar-on-Image Regression*

SIR is a statistical linear method used to study and analyse the relationship between a scalar outcome and two or three-dimensional predictor images under a single regression model [10, 14, 16, 19]. For each data point, $i = 1, \ldots, n$, we have

$$y_i = \mathbf{w}_i^T \boldsymbol{\mu} + \mathbf{x}_i^T \boldsymbol{\beta} + \epsilon_i, \quad \epsilon_i \overset{iid}{\sim} \mathrm{N}\left(0, \sigma^2\right), \tag{1}$$

where y_i is a scalar continuous outcome measure, $\mathbf{w}_i = (w_{i1}, \ldots, w_{iq})^T \in \mathbb{R}^q$ is a q-dimensional vector of covariates, and $\mathbf{x}_i = (x_{i1}, \ldots, x_{ip})^T \in \mathbb{R}^p$ is a p-dimensional image predictor. Each x_{ij} indicates the value of the image at a single pixel with spatial location $\mathbf{s}_j = (s_{j1}, s_{j2})^T \in \mathbb{R}^2$ for $j = 1, \ldots, p$. We define $\boldsymbol{\mu} = (\mu_1, \ldots, \mu_q)^T \in \mathbb{R}^q$ as a q-dimensional fixed effects vector and $\boldsymbol{\beta} = \left(\beta(\mathbf{s}_1), \ldots, \beta(\mathbf{s}_p)\right)^T$ (with $\beta_j := \beta(\mathbf{s}_j)$) as the spatially varying coefficient image described on the same lattice as $\mathbf{x}_i$. We model the high-dimensional $\boldsymbol{\beta}$ by spatially clustering the pixels into M regions and assuming common coefficients $\beta_1^*, \ldots, \beta_M^*$ within in each cluster, i.e. $\beta_j = \beta_m^*$ given the cluster label $z_j = m$. Thus, the prior on the coefficient image is decomposed into two parts: the random image partition model for spatially clustering the pixels and a shrinkage prior for the cluster-specific coefficients $\boldsymbol{\beta}^* = \left(\beta_1^*, \ldots, \beta_M^*\right)^T$. The SIR model in (1) can be extended for other types of responses through a generalized linear model framework (GLM) [23].

2.2 *Random Image Partition Model*

The image predictors are observed on a spatially structured coordinate system. Exchangeability is indeed no longer the proper assumption as the images contain covariate information, that we wish to leverage to improve model performance in this high-dimensional setting. To do so, we combine BNP random partition models, which avoid the need to prespecify the number of clusters, allowing it be determined and grow with the data, with a Potts-like spatial smoothness component [36]. Spatial random partition models in this direction are a growing research area, including Markov random field (MRF) with the product partition model (PPM) [32], with DP [29, 47], with Pitman-Yor process (PY)[21] and with mixture of finite mixtures (MFM) [13, 48]. Precisely, within the BNP framework, we focus on the class of Gibbs-type random partitions [1, 9, 20, 35], motivated by their comprise between tractable pre-

Table 1 Formulas of $V_p(M)$, $W_m(\phi)$ and terms of the predictive probability for assigning current cluster to either existing cluster or new cluster for DP, PY and MFM

	DP	PY	MFM
$V_p(M)$	$\frac{\Gamma(\alpha)\alpha^M}{\Gamma(\alpha+p)}$	$\frac{\Gamma(\alpha+1)\prod_{m=1}^{M-1}(\alpha+m\delta)}{\Gamma(\alpha+p)}$	$\sum_{l=1}^{\infty} \frac{\Gamma(\gamma l) l!}{\Gamma(\gamma l+p)(l-m)!} P_L(\cdot\|\lambda)$
$W_m(\phi)$	$\Gamma(\| C_m \|)$	$\frac{\Gamma(\|C_m\|-\delta)}{\Gamma(1-\delta)}$	$\frac{\Gamma(\|C_m\|+\gamma)}{\Gamma(\gamma)}$
Existing cluster	$\frac{\Gamma(\|C_m^{-A_o}\|+\|A_o\|)}{\Gamma(\|C_m^{-A_o}\|)}$	$\frac{\Gamma(\|C_m^{-A_o}\|+\|A_o\|-\delta)}{\Gamma(\|C_m^{-A_o}\|-\delta)}$	$\frac{\Gamma(\|C_m^{-A_o}\|+\|A_o\|+\gamma)}{\Gamma(\|C_m^{-A_o}\|+\gamma)}$
New cluster	$\alpha\Gamma(\|A_o\|)$	$(\alpha + \delta M^{-A_o})\frac{\Gamma(\|A_o\|-\delta)}{\Gamma(1-\delta)}$	$\frac{V_p(M^{-A_o}+1)}{V_p(M^{-A_o})}\frac{\Gamma(\|A_o\|+\gamma)}{\Gamma(\gamma)}$

Note that the predictive probabilities are stated up to a proportionality constant

dictive rules and richness of the predictive structure, including important cases, such as the DP [6], PY [33, 34], and MFM [25]. The Potts-Gibbs models induce a distribution on the partition $\pi_p = \{C_1, \ldots, C_M\}$ of p pixels into M nonempty, mutually exclusive, and exhaustive subsets $C_1, \ldots, C_M$ such that $\cup_{C \in \pi_p} C = \{1, \ldots, p\}$. The model can be summarised as:

$$\text{pr}(\pi_p) \propto \underbrace{\exp\left(\sum_{j\sim k, j<k} \upsilon_{jk}\mathbf{1}_{z_j=z_k}\right)}_{\text{Potts model}} \underbrace{\left(V_p(M)\prod_{m=1}^{M} W_m(\phi)\right)}_{\text{Gibbs-type random partition models}},$$

where $z_j \in \{1, \cdots, M\}$, $j \sim k$ means that j and k are neighbors, and $\mathbf{1}_{z_j=z_k}$ equals to 1 if j and k in the same cluster and 0 otherwise. In the following, we assume the spatial locations lie on a rectangular lattice with first-order neighbors and a common coupling parameter υ for all neighbor pairs; a higher value of υ encourages more spatial smoothness in the partition. We use the general notation ϕ to denote the parameters of the Gibbs-type partition models, and focus our study on three cases 1) DP with concentration parameter $\alpha > 0$; 2) PY with discount parameter $\delta \in [0, 1)$ and concentration parameter $\alpha > -\delta$; and 3) MFM with parameter $\gamma > 0$ (larger values encouraging more equally sized clusters) and a distribution $P_L(\cdot|\lambda)$ with parameter λ related to the prior on the number of clusters. The $\{V_p(M) : p \geq 1, 1 \leq M \leq p\}$ denotes the set of non-negative weights, which solves the backward recurrence relation $V_p(M) = (p - \delta M)V_{p+1}(M) + V_{p+1}(M + 1)$ with $V_1(1) = 1$. Table 1 describes the $V_p(M)$ and $W_m(\phi)$ for DP, PY and MFM models.

2.3 Shrinkage Prior

To identify relevant regions, we use heavy tailed priors for the unique values $(\beta_1^*, \ldots, \beta_M^*)$ of $(\beta(\mathbf{s}_1), \ldots, \beta(\mathbf{s}_p))$. Specifically, a t-shrinkage prior is used, motivated by its computational efficiency and nearly optimal contraction rate and selection consistency [41]:

$$\begin{aligned} \sigma^2 &\sim \text{IG}(a_\sigma, b_\sigma), \\ (\beta_m^*) \,|\sigma^2 &\sim t_\nu(s\sigma), \quad \text{for all } m = 1, \ldots, M, \end{aligned} \tag{2}$$

where $t_\nu(s\sigma)$ denotes t-distribution with degree of freedom ν and scale parameter $s\sigma$. For posterior inference, the t-distribution (2) is rewritten as a hierarchical inverse-gamma scaled Gaussian mixture,

$$\begin{aligned} \sigma^2 &\sim \text{IG}(a_\sigma, b_\sigma), \\ \eta_m^* &\sim \text{IG}(a_\eta, b_\eta), \\ (\beta_m^*) \,|\sigma^2, \eta_m^* &\sim N(0, \eta_m^* \sigma^2), \quad \text{for all } m = 1, \ldots, M, \end{aligned}$$

where a_η and b_η are the shape and scaling parameter of the mixing distribution for each η_m^* respectively with $\nu = 2a_\eta$ and $s = \sqrt{b_\eta / a_\eta}$.

3 Inference

We aim to infer the posterior distribution of the parameters based on the proposed Potts-Gibbs SIR model:

$$\begin{aligned} y_i \mid \boldsymbol{\mu}, \boldsymbol{\beta}^*, \pi_p, \sigma^2 &\sim \text{N}(\mathbf{w}_i^T \boldsymbol{\mu} + \mathbf{x}_i^{*T} \boldsymbol{\beta}^*, \sigma^2), \quad \text{for all } i = 1, \ldots, n, \\ \boldsymbol{\mu} \mid \sigma^2 &\sim \text{N}(\mathbf{m}_\mu, \sigma^2 \Sigma_\mu), \\ \boldsymbol{\beta}^* \mid \boldsymbol{\eta}^*, \sigma^2 &\sim \text{N}(\mathbf{0}_M, \sigma^2 \Sigma_{\beta^*}), \\ \sigma^2 &\sim \text{IG}(a_\sigma, b_\sigma), \\ \eta_m^* &\sim \text{IG}(a_\eta, b_\eta), \quad \text{for all } m = 1, \ldots, M, \\ \pi_p &\sim \text{Potts-Gibbs}(\upsilon, \phi), \end{aligned}$$

where $x_{im}^* = \sum_{j=1}^p x_{ij} \mathbf{1}(j \in C_m) / \sqrt{| C_m |}$ represents the total value, e.g. volume in the mth region of the image, $\mathbf{m}_\mu = (m_{\mu_1}, \ldots, m_{\mu_q})$, $\Sigma_\mu = \text{diag}(c_{\mu_1}, \ldots, c_{\mu_q})^T$, and $\Sigma_{\beta^*} = \text{diag}(\eta_1^*, \ldots, \eta_M^*)$. Note that when defining x_{im}^*, we rescale by the square root of cluster size , which is equivalent to rescaling the variance of β_m^* by the cluster size, encouraging more shrinkage for larger regions.

We develop a Gibbs sampler to simulate from the posterior with a generalized Swendsen-Wang (GSW) algorithm to draw samples from the Potts-Gibbs model. Poor mixing can be seen in single-site Gibbs sampling [8] due to the high correlation between the pixel labels. The SW algorithm [43] addresses this by forming nested clusters of neighbouring pixels, then updating all of the labels within a nested cluster to the same value. The generalisation of the technique for standard Potts models to generalised Potts-partition models is called GSW [47]. At each step of the algorithm, we proceed through the following steps:

1. Sample the image partition π_p given $\boldsymbol{\eta}^*$ and the data (with $\boldsymbol{\beta}^*, \boldsymbol{\mu}, \sigma^2$ marginalized). GSW is used to update simultaneously nested groups of pixels and hence improve the exploration of the posterior. The algorithm relies on the introduction of auxiliary binary bond variables, where $r_{jk} = 1$ if pixels j and k are bonded, otherwise 0. The bond variables define a partition of the pixels into nested clusters $A_1, \ldots, A_O$, where O denotes the number of nested clusters and each $A_o \subseteq C_m$ for some $m = 1, \ldots, M$. For each neighbor pair $j \sim k$ for $1 \leq j < k \leq p$, we sample the bond variables as follows, $r_{jk} \sim \text{Ber}\{1 - \exp(-\upsilon_{jk}\zeta_{jk}\mathbf{1}_{z_j = z_k})\}$, where we define $\zeta_{jk} = \kappa \exp\{-\tau d(\hat{\beta}_j, \hat{\beta}_k)\}$ with $\hat{\beta}_j$ denoting the estimated coefficient from univariate regression on the jth pixel and κ, τ are the tuning parameters of the GSW sampler. Notice that the algorithm reduces to single-site Gibbs when $\kappa = 0$, and recovers classical SW when $\kappa = 1$ and $\tau = 0$.
As we are dealing with non-conjugate priors, we update the cluster assignment by extending Gibbs sampling with the addition of auxiliary parameters, which is widely known as Algorithm 8 [28]. We denote by A_o the current nested cluster; $C_1^{-A_o}, \ldots, C_M^{-A_o}$ the clusters without nested cluster A_o; M^{-A_o} the number of distinct clusters excluding A_o and h the number of temporary auxiliary variables. For each nested cluster A_o, it is assigned to an existing cluster $m = 1, \ldots, M^{-A_o}$ or a new cluster $m = M^{-A_o} + 1, \ldots, M^{-A_o} + h$ with probability as follows,
$$\text{pr}(A_o \in C_m^{-A_o} \mid \cdots)$$
$$\propto \begin{cases} \frac{\Gamma(|C_m^{-A_o}| + |A_o| - \delta)}{\Gamma(|C_m^{-A_o}| - \delta)} \text{pr}\left(\mathbf{y} \mid \pi_p^{A_o \to m}, \boldsymbol{\eta}^*\right) \\ \prod_{\{(j,k) \mid j \in A_o, k \in C_m^{-A_o}, r_{jk} = 0\}} \exp\left\{\upsilon_{jk}(1 - \zeta_{jk})\right\}, & \text{for } C_m^{-A_o} \in \pi_p^{-A_o}, \\ \\ \frac{1}{h} \frac{V_p(M^{-A_o} + 1)}{V_p(M^{-A_o})} \frac{\Gamma(|A_o| - \delta)}{\Gamma(1 - \delta)} \text{pr}\left(\mathbf{y} \mid \pi_p^{A_o \to M+1}, \boldsymbol{\eta}^*\right), & \text{for new } C_m^{-A_o}; \end{cases}$$
where $\text{pr}\left(\mathbf{y} \mid \pi_p^{A_o \to m}, \boldsymbol{\eta}^*\right)$ and $\text{pr}\left(\mathbf{y} \mid \pi_p^{A_o \to M+1}, \boldsymbol{\eta}^*\right)$ denote the marginal likelihood of data obtained by moving A_o from its current cluster to existing clusters or newly created cluster respectively. Before updating the cluster assignments, we sample the nested clusters and compute the volume of each nested cluster for all images, with computational cost $\mathcal{O}(np)$. When updating the cluster assignments, the marginal likelihood dominates the computational cost, as it involves inversion and determinants of $(M + q) \times (M + q)$ matrices and updating the sufficient statistics for every nested cluster and every outer cluster allocation, i.e. the cost is $\mathcal{O}([[M + q]^3 + n[M + q]]OM)$.
2. Sample $\boldsymbol{\beta}^*, \boldsymbol{\mu}, \sigma^2$ jointly given the partition $\pi_p, \boldsymbol{\eta}^*$ and the data. Notationally, we reformulate $\tilde{\mathbf{x}}_i = (\mathbf{w}_i^T, \mathbf{x}_i^{*T})^T$ and $\tilde{\boldsymbol{\beta}} = (\boldsymbol{\mu}^T, \boldsymbol{\beta}^{*T})^T$. We define $\tilde{\mathbf{X}}$ be the matrix with rows equal to $\tilde{\mathbf{x}}_i^T$. The corresponding full conditional for $\tilde{\boldsymbol{\beta}}$ and σ^2 is
$$\sigma^2 \mid \cdots \sim \text{IG}(\hat{a}_\sigma, \hat{b}_\sigma),$$
$$\tilde{\boldsymbol{\beta}} \mid \sigma^2, \cdots \sim \text{N}(\hat{\mathbf{m}}_{\tilde{\beta}}, \sigma^2 \hat{\Sigma}_{\tilde{\beta}}),$$

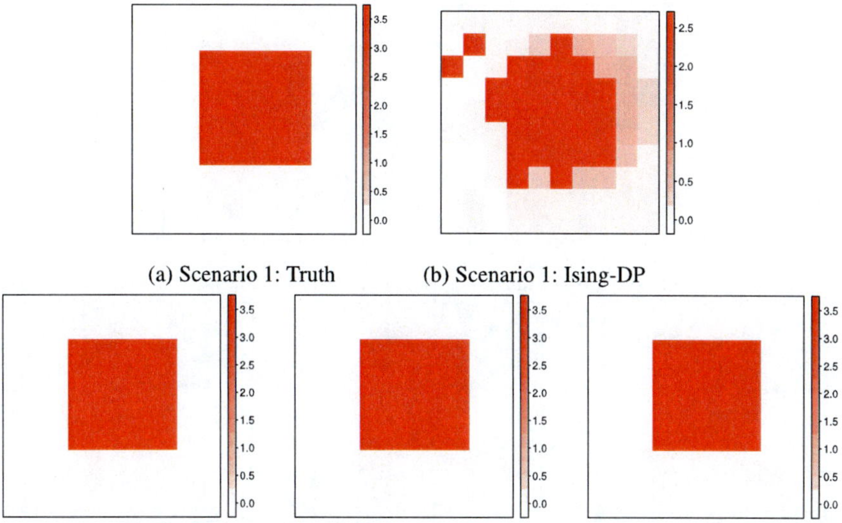

(a) Scenario 1: Truth (b) Scenario 1: Ising-DP

(c) Scenario 1: Potts-DP (d) Scenario 1: Potts-PY (e) Scenario 1: Potts-MFM

Fig. 1 Figures on the upper and bottom row showing the true and estimated coefficient matrix of the simulated data sets for scenario 1 under each model

where $\hat{\boldsymbol{\Sigma}}_{\tilde{\beta}} = (\boldsymbol{\Sigma}_{\tilde{\beta}}^{-1} + \tilde{\mathbf{X}}^T\tilde{\mathbf{X}})^{-1}$, $\hat{\mathbf{m}}_{\tilde{\beta}} = \hat{\boldsymbol{\Sigma}}_{\tilde{\beta}}(\boldsymbol{\Sigma}_{\tilde{\beta}}^{-1}\mathbf{m}_{\tilde{\beta}} + \tilde{\mathbf{X}}^T\mathbf{y})$, and $\mathrm{IG}(\hat{a}_\sigma, \hat{b}_\sigma)$ denotes the inverse-gamma distribution with updated shape $\hat{a}_\sigma = a_\sigma + n/2$ and scale $\hat{b}_\sigma = b_\sigma + [\mathbf{m}_{\tilde{\beta}}^T\boldsymbol{\Sigma}_{\tilde{\beta}}^{-1}\mathbf{m}_{\tilde{\beta}} + \mathbf{y}^T\mathbf{y} - \hat{\mathbf{m}}_{\tilde{\beta}}^T\hat{\boldsymbol{\Sigma}}_{\tilde{\beta}}^{-1}\hat{\mathbf{m}}_{\tilde{\beta}}]/2$.

3. Sample $\boldsymbol{\eta}^*$ given $\boldsymbol{\beta}^*$. The corresponding full conditional for each η_m^* is an inverse-gamma distribution with updated shape $\hat{a}_\eta = a_\eta + 1/2$ and scale $\hat{b}_\eta = b_\eta + (\beta_m^*)^2/(2\sigma^2)$:

$$\eta_m^* \mid \cdots \sim \mathrm{IG}(\hat{a}_\eta, \hat{b}_\eta), \quad \text{for } m = 1, \ldots, M.$$

4 Numerical Studies

We study through simulations the performance of the proposed model and compare it with Ising-DP [19]. We consider 2D images in this simulation. The $n = 300$ images are simulated on a two dimensional grid of size 10×10, with spatial locations $\mathbf{s}_j = (s_{j1}, s_{j2}) \in \mathbf{R}^2$ for $1 \le s_{j1}, s_{j2} \le 10$. For simplicity's sake, we include an intercept but do not consider others covariates, $\mathbf{w}_i$. We concentrate on the two simulation scenarios with true $M = 2$ and $M = 5$ as shown in Figs. 1 and 2. For each experiment, we summarise the posterior of the clustering structure of the data sets by minimising the posterior expected Variation of Information (VI) [45].

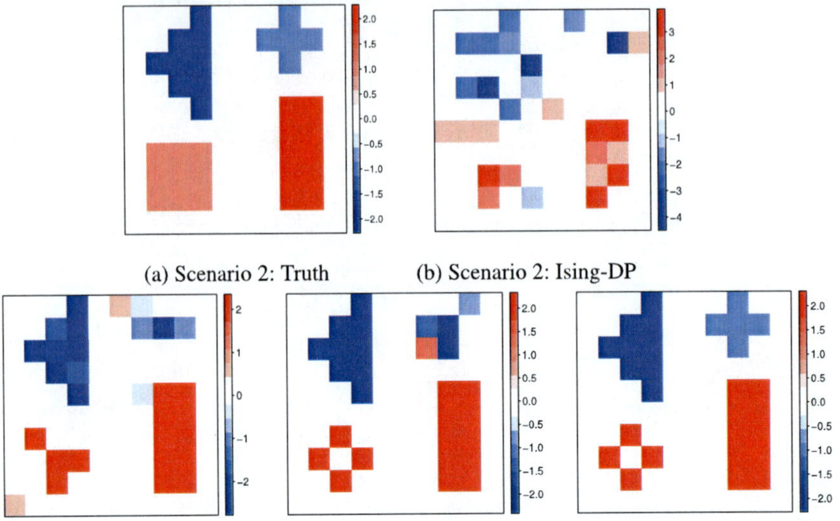

(a) Scenario 2: Truth (b) Scenario 2: Ising-DP

(c) Scenario 2: Potts-DP (d) Scenario 2: Potts-PY (e) Scenario 2: Potts-MFM

Fig. 2 Figures on the upper and bottom row showing the true and estimated coefficient matrix of the simulated data sets for scenario 2 under each model

The Potts-Gibbs models can detect correctly the cluster structure under scenario 1 (Fig. 1). The Potts-Gibbs models are also capable of capturing and identifying the more complex cluster structure underlying the data for scenario 2 (Fig. 2) with the ARI 0.621–0.830 (Table 2). On the contrary, Ising-DP has failed terribly to recover the cluster structure for scenario 2, as illustrated in Fig. 2. It is observed that under the Potts-Gibbs models, most of the resultant clusters are spatially proximal, while under Ising-DP, the clusters are dispersed throughout the image. By taking into consideration spatial dependence in the random partition model via the Potts-Gibbs models, the proposed models produce spatially aware clustering and thus improve the predictions.

DP has a concentration parameter α, with larger values encouraging more new clusters and a rich-get-richer property that favours allocation to larger clusters. The PY has an additional discount parameter $\delta \in [0, 1)$ that helps to mitigate the rich-get-richer property and phase transition of the Potts model. The MFM has a parameter γ, with larger values encouraging more equal-sized clusters and helping to avoid phase transition of the Potts model, as well as additional parameters λ which are related to the prior on the number of clusters.

Table 2 Mean, standard deviation (in parentheses) and highest posterior density (HPD) interval of the posterior of adjusted Rand index (ARI), variation information (VI), mean squared error (MSE), mean squared prediction error (MSPE), and number of clusters for each scenario under each model

	Model	Scenario	Mean	HPD (95%)
ARI	Potts-DP	1	**1.0** (0.004)	(1.0, 1.0)
	Potts-PY		**1.0** (0.004)	(1.0, 1.0)
	Potts-MFM		**0.999** (0.007)	(1.0, 1.0)
	Ising-DP		0.307 (0.079)	(0.152, 0.464)
	Potts-DP	2	**0.621** (0.060)	(0.472, 0.684)
	Potts-PY		**0.713** (0.050)	(0.607, 0.818)
	Potts-MFM		**0.830**(0.036)	(0.756, 0.869)
	Ising-DP		0.038 (0.021)	(-0.001, 0.078)
VI	Potts-DP	1	**0.001** (0.010)	(2.22e-16, 2.22e-16)
	Potts-PY		**0.001** (0.009)	(2.220e-16, 2.220e-16)
	Potts-MFM		**0.001** (0.014)	(2.220e-16, 2.220e-16)
	Ising-DP		1.386 (0.154)	(1.083, 1.680)
	Potts-DP	2	**1.160** (0.211)	(0.902, 1.548)
	Potts-PY		**1.006** (0.147)	(0.640, 1.299)
	Potts-MFM		**0.599** (0.133)	(0.432, 0.866)
	Ising-DP		3.990 (0.159)	(3.691, 4.290)
MSE	Potts-DP	1	**1.33e-4** (5.59e-4)	(3.97e-9, 2.67e-4)
	Potts-PY		**1.03e-4** (8.73e-5)	(1.58e-7, 2.57e-4)
	Potts-MFM		**1.01e-4** (8.37e-5)	(4.21e-7, 2.66e-4)
	Ising-DP		0.807 (0.011)	(0.790, 0.828)
	Potts-DP	2	**0.246** (0.064)	(0.141, 0.374)
	Potts-PY		**0.157** (0.035)	(0.094, 0.224)
	Potts-MFM		**0.093** (0.014)	(0.079, 0.125)
	Ising-DP		0.980 (0.025)	(0.942, 1.020)
MSPE	Potts-DP	1	**4.215** (0.057)	(4.152, 4.317)
	Potts-PY		**4.213** (0.052)	(4.138, 4.311)
	Potts-MFM		**4.209** (0.052)	(4.136, 4.314)
	Ising-DP		145.912 (10.051)	(129.142, 165.950)
	Potts-DP	2	7.754 (2.653)	(3.175, 13.356)
	Potts-PY		**0.868**(0.168)	(0.669, 1.189)
	Potts-MFM		**0.850** (0.122)	(0.677, 1.108)
	Ising-DP		3.641 (0.526)	(2.766, 4.857)
Number of clusters	Potts-DP	1	2.019 (0.138)	(2.0, 2.0)
	Potts-PY		2.015 (0.122)	(2.0, 2.0)
	Potts-MFM		2.007 (0.081)	(2.0, 2.0)
	Ising-DP		4.575 (1.340)	(2.0, 7.0)
	Potts-DP	2	6.722 (0.901)	(5.0, 8.0)
	Potts-PY		6.882 (1.090)	(5.0, 9.0)
	Potts-MFM		5.232 (0.475)	(5.0, 6.0)
	Ising-DP		15.542 (1.554)	(13.0, 18.0)

5 Conclusion

We have developed novel Bayesian scalar-on-image regression models to extract interpretable features from the image by clustering and leveraging the spatial coordinates of the pixels/voxels. To encourage groups representing spatially contiguous regions, we incorporate the spatial information directly in the prior for the random partition through Potts-Gibbs random partition models. We have shown the potential of Potts-Gibbs models in detecting the correct cluster structure on simulated data sets. In our experiments, the hyperparameters of the Potts-Gibbs model were determined via a simple grid search on selected combinations of hyperparameters. However, future work will consist of investigating the influence of the various parameters inherent to the model and guidelines and tools to determine hyperparameters. The model will then be applied to real images, e.g. neuroimages. Motivated by examining and identifying brain regions of interest in Alzheimer's disease, we will use MRI images obtained from the Alzheimer's Disease Neuroimaging Initiative (ADNI) database (www.adni-info.org). The proposed SIR model will be extended to classification problems through the GLM framework.

References

1. Cerquetti, A.: Generalized Chinese restaurant construction of exchangeable Gibbs partitions and related results. arXiv:0805.3853 (2008)
2. Craddock, R.C., Holtzheimer, P.E., III., Hu, X.P., Mayberg, H.S.: Disease state prediction from resting state functional connectivity. Magn. Reson. Med. **62**(6), 1619–1628 (2009)
3. Davatzikos, C., Shen, D., Gur, R.C., Wu, X., Liu, D., Fan, Y., Hughett, P., Turetsky, B.I., Gur, R.E.: Whole-brain morphometric study of schizophrenia revealing a spatially complex set of focal abnormalities. Arch. Gen. Psychiatry **62**(11), 1218–1227 (2005)
4. Debois, D., Ongena, M., Cawoy, H., De Pauw, E.: MALDI-FTICR MS imaging as a powerful tool to identify Paenibacillus antibiotics involved in the inhibition of plant pathogens. J. Am. Soc. Mass Spectrom. **24**(8), 1202–1213 (2013)
5. Fan, Y., Resnick, S.M., Wu, X., Davatzikos, C.: Structural and functional biomarkers of prodromal Alzheimer's disease: a high-dimensional pattern classification study. NeuroImage **41**(2), 277–285 (2008)
6. Ferguson, T.S.: A Bayesian analysis of some nonparametric problems. Ann. Stat. **1**, 209–230 (1973)
7. Ferwerda, B., Schedl, M., Tkalcic, M.: Using instagram picture features to predict users' personality. In: International Conference on Multimedia Modeling. Springer (2016)
8. Geman, S., Geman, D.: Stochastic relaxation, Gibbs distributions, and the Bayesian restoration of images. IEEE Trans. Pattern Anal. Mach. Intell. **6**, 721–741 (1984)
9. Gnedin, A., Pitman, J.: Exchangeable Gibbs partitions and Stirling triangles. J. Math. Sci. **138**(3), 5674–5685 (2006)
10. Goldsmith, J., Huang, L., Crainiceanu, C.M.: Smooth scalar-on-image regression via spatial Bayesian variable selection. J. Comput. Graph Stat. **23**(1), 46–64 (2014)
11. Gundlach-Graham, A., Burger, M., Allner, S., Schwarz, G., Wang, H.A.O., Gyr, L., Grolimund, D., Hattendorf, B., Günther, D.: High-speed, high-resolution, multielemental laser ablation-inductively coupled plasma-time-of-flight mass spectrometry imaging: Part i. instrumentation and two-dimensional imaging of geological samples. Anal. Chem. **87**(16), 8250–8258 (2015)

12. Henderson, J.V., Storeygard, A., Weil, D.N.: Measuring economic growth from outer space. National Bureau of Economic Research, Cambridge, Mass (2009)
13. Hu, G., Geng, J., Xue, Y., Sang, H.: Bayesian spatial homogeneity pursuit of functional data: an application to the U.S. income distribution. arXiv:2002.06663 (2020)
14. Huang, L., Goldsmith, J., Reiss, P.T., Reich, D.S., Crainiceanu, C.M.: Bayesian scalar-on-image regression with application to association between intracranial DTI and cognitive outcomes. NeuroImage **83**, 210–223 (2013)
15. Hum, N.J., Chamberlin, P.E., Hambright, B.L., Portwood, A.C., Schat, A.C., Bevan, J.L.: A picture is worth a thousand words: a content analysis of Facebook profile photographs. Comput. Hum. Behav. **27**(5), 1828–1833 (2011)
16. Kang, J., Reich, B.J., Staicu, A.M.: Scalar-on-image regression via the soft-thresholded Gaussian process. Biometrika **105**(1), 165–184 (2018)
17. Kim, Y., Kim, J.H.: Using computer vision techniques on instagram to link users' personalities and genders to the features of their photos: an exploratory study. Inf. Process. Manage. **54**(6), 1101–1114 (2018)
18. Lee, K., Cao, X.: Bayesian group selection in logistic regression with application to MRI data analysis. Biometrics **77**(2), 391–400 (2021)
19. Li, F., Zhang, T., Wang, Q., Gonzalez, M.Z., Maresh, E.L., Coan, J.A.: Spatial Bayesian variable selection and grouping for high-dimensional scalar-on-image regression. Ann. Appl. Stat. **9**(2), 687–713 (2015)
20. Lijoi, A., Prünster, I.: Models beyond the Dirichlet process. In: Bayesian Nonparametrics (2010)
21. Lü, H., Arbel, J., Forbes, F.: Bayesian nonparametric priors for hidden Markov random fields. Stat. Comput. **30**(4), 1015–1035 (2020)
22. Maloof, K.A., Reinders, A.N., Tucker, K.R.: Applications of mass spectrometry imaging in the environmental sciences. Curr. Opin. Environ. Sci. Health. **18**, 54–62 (2020)
23. McCullagh, P., Nelder, J.A.: Generalized linear models. Routledge (2019)
24. Mehrotra, S., Maity, A.: Simultaneous variable selection, clustering, and smoothing in function-on-scalar regression. Can. J, Stat (2021)
25. Miller, J.W., Harrison, M.T.: Mixture models with a prior on the number of components. J. Am. Stat. Assoc. **113**(521), 340–356 (2018)
26. Naik, N., Kominers, S.D., Raskar, R., Glaeser, E.L., Hidalgo, C.A.: Computer vision uncovers predictors of physical urban change. Proc. Natl. Acad. Sci. U.S.A. **114**(29), 7571–7576 (2017)
27. Naik, N., Raskar, R., Hidalgo, C.A.: Cities are physical too: using computer vision to measure the quality and impact of urban appearance. Am. Econ. Rev. **106**(5), 128–132 (2016)
28. Neal, R.M.: Markov chain sampling methods for Dirichlet process mixture models. J. Comput. Graph Stat. **9**(2), 249–265 (2000)
29. Orbanz, P., Buhmann, J.M.: Nonparametric Bayesian image segmentation. Int. J. Comput. Vis. **77**(1–3), 25–45 (2007)
30. O'Neill, S.J.: Image matters: climate change imagery in US. UK and Australian newspapers. Geoforum **49**, 10–19 (2013)
31. O'Neill, S.J., Boykoff, M., Niemeyer, S., Day, S.A.: On the use of imagery for climate change engagement. Glob. Environ. Change **23**(2), 413–421 (2013)
32. Pan, T., Hu, G., Shen, W.: Identifying latent groups in spatial panel data using a Markov random field constrained product partition model. arXiv:2012.10541 (2020)
33. Perman, M., Pitman, J., Yor, M.: Size-biased sampling of Poisson point processes and excursions. Probab. Theory Relat. Fields **92**(1), 21–39 (1992)
34. Pitman, J.: Some developments of the Blackwell-Macqueen urn scheme. Lect. Notes-Monograph Ser. **30**, 245–267 (1996)
35. Pitman, J.: Lecture Notes in Mathematics. Springer (2006)
36. Potts, R.B., Domb, C.: Some generalized order-disorder transformations. Math. Proc. Cambridge Philos. Soc. **48**(1), 106 (1952). https://doi.org/10.1017/S0305004100027419
37. Reiss, P., Mennes, M., Petkova, E., Huang, L., Hoptman, M., Biswal, B., Colcombe, S., Zuo, X., Milham, M.: Extracting information from functional connectivity maps via function-on-scalar regression. NeuroImage **56**, 140–148 (2011)

38. Samany, N.N.: Automatic landmark extraction from geo-tagged social media photos using deep neural network. Cities **93**, 1–12 (2019)
39. Shi, J., Lepore, N., Gutman, B., Thompson, P., Baxter, L., Caselli, R., Wang, Y.: Genetic influence of apolipoprotein E4 genotype on hippocampal morphometry: an N = 725 surface-based Alzheimer's disease neuroimaging initiative study. Hum. Brain Mapp. **35**(8), 3903–3918 (2014)
40. Smith, M., Fahrmeir, L.: Spatial Bayesian variable selection with application to functional magnetic resonance imaging. J. Am. Stat. Assoc. **102**(478), 417–431 (2007)
41. Song, Q., Liang, F.: Nearly optimal Bayesian shrinkage for high dimensional regression. arXiv:1712.08964 (2017)
42. Sun, D., van Erp, T.G., Thompson, P.M., Bearden, C.E., Daley, M., Kushan, L., Hardt, M.E., Nuechterlein, K.H., Toga, A.W., Cannon, T.D.: Elucidating a magnetic resonance imaging-based neuroanatomic biomarker for psychosis: Classification analysis using probabilistic brain atlas and machine learning algorithms. Biol. Psychiatry (1969) **66**(11), 1055–1060 (2009)
43. Swendsen, R.H., Wang, J.S.: Nonuniversal critical dynamics in Monte Carlo simulations. Phys. Rev. Lett. **58**(2), 86–88 (1987)
44. Van Walderveen, M., Kamphorst, W., Scheltens, P., Van Waesberghe, J., Ravid, R., Valk, J., Polman, C., Barkhof, F.: Histopathologic correlate of hypointense lesions on T1-weighted spin-echo MRI in multiple sclerosis. Neurology **50**(5), 1282–1288 (1998)
45. Wade, S., Ghahramani, Z.: Bayesian cluster analysis: Point estimation and credible balls (with discussion). Bayesian Anal. **13**(2), 559–626 (2018)
46. Wang, X., Zhu, H., Initiative, A.D.N.: Generalized scalar-on-image regression models via total variation. J. Am. Stat. Assoc. **112**(519), 1156–1168 (2017)
47. Xu, R.Y.D., Caron, F., Doucet, A.: Bayesian nonparametric image segmentation using a generalized Swendsen-Wang algorithm. arXiv:1602.03048 (2016)
48. Zhao, P., Yang, H.C., Dey, D.K., Hu, G.: Bayesian spatial homogeneity pursuit regression for count value data. arXiv:2002.06678 (2020)

Block Structured Graph Priors in Gaussian Graphical Models

Alessandro Colombi

Abstract Gaussian graphical models are a powerful statistical tool to describe the concept of conditional independence between variables through a map between a graph and the family of multivariate normal models. The structure of the graph is unknown and has to be learned from the data. Inference is carried out in a Bayesian framework: thus, the structure of the precision matrix is constrained by the graph through a G-Wishart prior distribution. In this work we first introduce a prior distribution to impose a block structure in the adjacency matrix of the graph. Then we develop a Double Reversible Jump Monte Carlo Markov chain that avoids any G-Wishart normalizing constant calculation when comparing graphical models. The novelty of this procedure is that it looks for block structured graphs, hence proposing moves that add or remove not just a single link but an entire group of them.

Keywords Bayesian statistics · Double reversible jump · G-Wishart prior

1 Introduction

The increasing capacity of human beings of collecting large amount of data gave rise to the need of developing models to study how variables interact with one another. Benefits of such discoveries are well known, for example in clinical and genetic applications it is useful to understand how risk factors are related so that patient-specific therapies may be planned. See [5, 9, 27] for cancer applications. The same reasoning applies to problems in economics, for example [25] studied the interconnectedness of credit risk.

Probabilistic graphical modeling is a possible approach to the task of studying the dependence structure among a set of variables. It relies on the concept of conditional independence between variables that is described through a map between a graph and a family of multivariate probability models. When such a family of probabilities

A. Colombi (✉)
Universitá di Milano-Bicocca, Milan, Italy
e-mail: a.colombi10@campus.unimib.it

R. Argiento et al. (eds.), *New Frontiers in Bayesian Statistics*, Springer Proceedings in Mathematics & Statistics 405, https://doi.org/10.1007/978-3-031-16427-9_6

is chosen to be Gaussian, those models are known as Gaussian graphical models [12]. This is the choice made throughout the paper, which is the most common in the literature.

Let $\boldsymbol{X}$ be a p-random vector distributed as $N_p(\mathbf{0}, \boldsymbol{\Sigma})$. $\boldsymbol{\Sigma}$ is the covariance matrix and we assume $\boldsymbol{X}$ to be centered without loss of generality. Let $G = (V, E)$ be an undirected graph, where $V = \{1, \ldots, p\}$ is the set of nodes and E is the set of undirected edges. $\boldsymbol{X}$ is said to be Markov with respect to G if, for any edge (i, j) that does not belong to E, the i-th and j-th variables are conditionally independent given all the others. Moreover, under the normality assumption, the conditional independence relationship between variables can be represented in terms of the null elements of the precision matrix $\boldsymbol{K} = \boldsymbol{\Sigma}^{-1}$. Therefore the following equivalence provides an interpretation of the graph

$$X_i \perp\!\!\!\perp X_j \mid \boldsymbol{X}_{-(ij)} \iff (i, j) \notin E \iff k_{ij} = 0, \tag{1}$$

where $\boldsymbol{X}_{-(ij)}$ is the random vector containing all elements in $\boldsymbol{X}$ except the i-th and the j-th. Each node is associated to one of the variables of interest and its links describe the structure of the non-zero elements of the precision matrix. The absence of a link between two vertices means that the two corresponding variables are conditionally independent, given all the others. Usually, G is unknown and it is the goal of the statistical inference, along with $\boldsymbol{K}$. Such a process is also known as structural learning. In a Bayesian framework, we set a G-Wishart prior distribution for the precision matrix $\boldsymbol{K}$ [1, 20] , which is attractive as it is conjugate to the likelihood. Since the graph G is considered to be a random variable having values in the space $\mathcal{G}$ of all possible undirected graphs with p nodes, we need to specify a prior on it. A common practice is to choose an uniform distribution over $\mathcal{G}$. This is appealing for its simplicity but it assigns most of its mass to graphs with a "medium" number of edges [9]. On the other hand, it is known that an undirected graph is uniquely identified by its set of edges $\mathcal{E}$. Therefore it is simpler to define a prior on $\mathcal{E}$, which then naturally induces a prior over $\mathcal{G}$. In this setting, the most natural choice is to assign independent Bernoulli priors to each link. The Bernoulli parameters θ could be different from edge to edge, but one usually assigns a common value. For example, [9] suggested to choose $\theta = 2/(p-1)$ to induce more sparsity in the graph. Scott and Carvalho [21] placed instead a Beta hyperprior on that parameter, a solution known as multiplicity correction prior. Similarly, [22] described a multivariate Bernoulli distribution where edges are not necessarily independent. A common feature of previously described priors is that they are non-informative. The only type of prior information they can include in the model is the expected sparsity.

In this work we propose a prior for the graph that aims to be informative, according to the prior information available for the application at hand. Since the graph describes the conditional dependence structure of variables involved in complex and high-dimensional phenomena, it is unrealistic to assume that prior knowledge is available for one-to-one relationships between the observed quantities. It is instead more reasonable to assume that variables may be grouped in smaller subsets. This is common in biological application where the groups may be families of bacteria [17], or genomics where groups of genes are known to be part of a common process. Also in market basket analysis products and customers can easily be grouped; see, for instance [6].

We propose a class of priors, called *block graph priors*, that encodes such information and imposes a block structure in the adjacency matrix that describes the graph. We allow variables in different groups only to be fully connected or not connected at all. Therefore, the goal is no longer in looking for all possible relationships between nodes but on deriving the underlying pattern between groups.

We introduce a Reversible Jump sampler that leverages the structure induced by our new prior. In particular, we generalize the procedure by Lenkoski [13]. The resulting method is called Block Double Reversible Jump (BDRJ for short). Its main feature is that it modifies, at each step of the chain, an entire block of links to guarantee a block structure that is always compatible with our hypotheses.

The remainder of the paper is organized as follows. Section 2 introduces the block structured graph priors and Sect. 3 provides the sampling strategy. In Sect. 4 we present a simulation study along with a comparison against an existing approach. Finally, we conclude with a brief discussion in Sect. 5.

2 Block Structured Graph Priors

The starting point for our proposed model is that we assume the p observed variables to be grouped, a priori, in M mutually exclusive groups. Each group has cardinality n_i and $\sum_{i=1}^{M} n_i = p$. We admit the possibility of having some $n_i = 1$, as long as $M < p$. Groups whose cardinality is equal to one are called *singletons*.

We aim to study relationships between groups of variables. Therefore the usual graph representation $G = (V, E)$, where V is the set of nodes and E is set of links, is redundant. Indeed we assume that groups are given and links have to satisfy a precise block structure. As a consequence, we synthesize those information by defining a new space of undirected graphs whose nodes represent the chosen groups of variables and links represent the structure of relationships between them. Namely, let $V_B = \{B_1, \ldots, B_M\}$ be a partition of V in M groups that are available a priori. Then we define $G_B = (V_B, E_B)$ to be an undirected graph whose nodes are the sets $B_k, k = 1, \ldots, M$ and that allows for self-loops if $n_k > 1$. Namely,

$$E_B \subset \mathcal{E}_B = \Big\{(l, m) | l, m \in V_B, \quad \wedge\, l < m, \quad (l, l) | l \in V_B, \quad \wedge\, n_l > 1\Big\}. \tag{2}$$

In graph theory, graphs that have self-loops are called *multigraphs*. Finally, let $\mathcal{G}_B$ be the set of all possible multigraphs G_B having V_B as set of nodes. In the following, we want to clarify the relationship between this space and $\mathcal{G}$.

Consider $G_B \in \mathcal{G}_B$ and $G \in \mathcal{G}$. By definition, the set of nodes of the first multigraph is obtained by grouping together the nodes of the second graph. What about

Fig. 1 The map from multigraph $G_B \in \mathcal{G}_B$ (left) to its block structured form $G \in \mathcal{B}$ (right)

the set of edges? Is there any relation between the two sets? The following map defines a relationship between them. Let $\rho : \mathcal{G}_B \to \mathcal{G}$, such that $G_B = (V_B, E_B) \mapsto G = (V, E)$ by the following transformations

$$\begin{aligned} &V = \{B_{l,h}, l = 1, \dots n_h, h = 1, \dots M\} = \{1, \dots, p\} \\ &\text{if } (l, m) \in E_B \Rightarrow \ (i, j) \in E \ \forall i \in B_l, \quad \forall j \in B_m \\ &\text{if } (l, m) \notin E_B \Rightarrow \ (i, j) \notin E \ \forall i \in B_l, \quad \forall j \in B_m \end{aligned} \tag{3}$$

A visual representation of this mapping is given in Fig. 1. Once ρ is set we are able to associate each G_B in $\mathcal{G}_B$ to one and only one G in $\mathcal{G}$, since ρ is clearly injective. We refer to G_B as the multigraph form of G.

Nevertheless, ρ is not surjective which implies that there are graphs that do not have a representative in $\mathcal{G}_B$. Indeed, only those graphs with a particular block structure can be represented in a multigraph form. A non surjective map is the key ingredient to define a subset of $\mathcal{G}$ of block structured graphs that satisfy our modelling assumptions. Let us consider the image of ρ, denoted by $\mathcal{B}$. It is the subset of $\mathcal{G}$ containing all the graphs having p nodes and a block structure consistent with V_B. Moreover, $\rho : \mathcal{G}_B \to \mathcal{B}$ is a bijection, which means that every graph $G \in \mathcal{B}$ is associated to its representative $G_B \in \mathcal{G}_B$ via ρ^{-1}. We say that $G \in \mathcal{B}$ is the block graph representation of the multigraph $G_B \in \mathcal{G}_B$. This synthesised representation of block graphs allows us to work in a space where we can use standard tools of graphical analysis. In a different setting, [4] adopts a similar approach to model the conditional dependence across Markov processes.

In particular, such a representation allows us to introduce a class of priors that encodes the knowledge about the partition of the nodes. We place zero mass probability on all those graphs that belong to $\mathcal{G} \backslash \mathcal{B}$, which is the set of all those graphs that do not satisfy our block structure constraint. Then, we place a standard prior, say $\pi_B(\cdot)$, over $\mathcal{G}_B$, which is possible as it is a space of undirected multigraphs where links can be considered to be independent with one another. Finally, we map the results in $\mathcal{B}$ using ρ^{-1}. Namely

$$\pi(G) \propto \begin{cases} \pi_B\left(\rho^{-1}(G)\right), & \text{if } G \in \mathcal{B} \\ 0, & \text{if } G \in \mathcal{G} \backslash \mathcal{B}. \end{cases} \tag{4}$$

We refer to those priors as *block graph priors*. In this work, we consider a *block-Bernoulli prior*, $\pi(G)$, that is obtained by applying (4) to $\pi_B(G_B) = \theta^{|E_B|}(1-\theta)^{\binom{M}{2}-|E_B|}$, that is the Bernoulli prior where each link has prior probability of inclusion θ, which is fixed a priori. The reasoning used to define block priors is similar to priors described in [7]. However, in this case we are not limiting the learning of the graph to the class of the decomposable ones but to block structured graphs. Moreover, this limitation is not due to computational limitations but because prior knowledge is available. In the next section we present a method to learn such a structure. In principle, one can still apply such prior to limit the analysis to the class of decomposable block graphs to exploit their properties. However, in this work we do not make such assumption and we present a method that is valid also for non-decomposable graphs.

3 Sampling Strategy

One of the difficulties in the development of efficient methods for structural learning is the presence of the G-Wishart prior distribution. Given a random matrix $\boldsymbol{K}$, we say that $\boldsymbol{K}|G, b, D \sim$ G-Wishart (b, D) if its density is

$$P(\boldsymbol{K} \mid G, b, D) = I_G(b, D)^{-1} |\boldsymbol{K}|^{\frac{b-2}{2}} \exp\left\{-\frac{1}{2} tr(\boldsymbol{K}D)\right\} \mathbb{1}_{\mathbb{P}_G}(\boldsymbol{K}), \tag{5}$$

where b and D are fixed hyperparameters, $\mathbb{P}_G$ is the space of all $p \times p$ symmetric and positive definite matrices whose null elements are associated to links absent in graph G and

$$I_G(b, D) = \int_{\mathbb{P}_G} |\boldsymbol{K}|^{\frac{b-2}{2}} \exp\left\{-\frac{1}{2} tr(\boldsymbol{K}D)\right\} d\boldsymbol{K}, \tag{6}$$

is an intractable normalizing constant. Numerical methods to approximate such a constant [1] are unstable in high dimensional problems [9, 15]. Several techniques that avoid any calculation of $I_G(b, D)$ are available in the literature, but an exhaustive review of them is beyond the goals of this work. In the following, we limit ourselves to present how our proposed method, called Block Double Reversible Jump (BDRJ for short). It generalizes the procedure by Lenkoski [13] to get a Reversible Jump chain defined over the joint space of graph and precision matrix that visits only the subspace $\mathcal{B}$ of block structured graphs. Note that if one is interested only in decomposable graph models, the normalizing constant $I_G(b, D)$ can be computed explicitly and it would be enough to use a standard Metropolis-Hastings algorithm without resorting to the usage of the Reversible Jump technique presented in the remaining part of this paper.

We denote the current state of the chain by $(\boldsymbol{K}^{[s]}, G^{[s]})$, with $\boldsymbol{K}^{[s]} \in \mathbb{P}_{G^{[s]}}$. The proposed state $(\boldsymbol{K}', G')$ is constructed in two subsequent steps; in Sect. 3.1 we describe the proposal for the new graph G' and then in Sect. 3.2 we discuss how to get the proposed precision matrix $\boldsymbol{K}' \in \mathbb{P}_{G'}$. Once that $(\boldsymbol{K}', G')$ has been drawn, we accept or reject the whole state with a Metropolis-Hastings step.

3.1 Construction of Proposed Graph G′

A common factor in most of the existing MCMC methods for graphical models is to set up chains such that the proposed graph $G' = (V, E')$ belongs to the one-edge-away neighbourhood of G. Namely, $nbd_p(G) = nbd_p^+(G) \cup nbd_p^-(G)$ where $nbd_p^+(G)$ and $nbd_p^-(G)$ are the sets of undirected graphs having p nodes that can be obtained by adding, or removing, an edge to $G \in \mathcal{G}$, respectively. A step in the Markov chain that selects $G' \in nbd_p(G^{[s]})$ is said to be a local move.

The proposed BDRJ approach is innovative because we derive moves that modifies an entire block of links, not just a single one. In other words, our moves are local in $\mathcal{G}_B$ but not in $\mathcal{G}$. Suppose $G^{[s]} \in \mathcal{B}$, we propose a new graph $G' \in \mathcal{B}$ by first drawing its multigraph representation $G'_B \in \mathcal{G}_B$ from

$$q(G'_B|G^{[s]}) = \frac{1}{2}\text{Unif}\left(nbd_M^{\mathcal{B},+}\left(\rho^{-1}(G^{[s]})\right)\right) + \frac{1}{2}\text{Unif}\left(nbd_M^{\mathcal{B},-}\left(\rho^{-1}(G^{[s]})\right)\right), \tag{7}$$

where $nbd_M^{\mathcal{B}}(G_B^{[s]})$ is the one-edge-away neighbourhood of $G_B^{[s]} = \rho^{-1}(G^{[s]})$ with respect to the space of multigraphs $\mathcal{G}_B$. Addition and removal moves are chosen with the same probability. Given this choice, $q(G'_B|G^{[s]})$ chooses, with uniform probability, which link is to be added (or removed). Finally ρ is applied once again to map the resulting multigraph back in $\mathcal{B}$ to obtain G', i.e. setting $G' = \rho\left(G'_B\right)$. A closer look at (7) reveals how our multigraph representation allows us to use standard tools of structural learning in the space $\mathcal{G}_B$ to get non-standard proposal in the usual space $\mathcal{G}$.

3.2 Construction of Proposed Precision Matrix K′

Once that the graph is selected, we need to specify a method to construct a proposed precision matrix $\boldsymbol{K}'$ that satisfies the constraints imposed by the new graph. The method by Wang and Li [26] based on the partial analytical structure of the G-Wishart appears to be an efficient choice. However, it strongly relies on the possibility of writing down an explicit formula for the full conditional of the elements of $\boldsymbol{K}$. Such results, presented in [20], can be handled in practice only if at each step of the graph only one link of the graph is modified. Instead, the proposal distribution presented in Sect. 3.1 modifies an arbitrary number of links. Hence, it is complicated, if possible at all, to generalize the method by Wang and Li [26] to such framework. As a consequence, we rely on a generalization of the Reversible Jump mechanism by Lenkoski [13]. The idea is that is it possible to guarantee the positive definiteness of

$\boldsymbol{K}'$ and the zero constraints imposed by G' just by working on the Cholesky decomposition matrix $\boldsymbol{\Phi}^{[s]}$ of $\boldsymbol{K}^{[s]}$. Indeed, [20] and [1] showed that the zero constraints imposed by $G^{[s]}$ on the off-diagonal elements of $\boldsymbol{K}^{[s]}$ induce a precise structure and properties on $\boldsymbol{\Phi}^{[s]}$. Let $\nu(G^{[s]}) = \{(i, j) \mid i, j \in V, i = j \text{ or } (i, j) \in E^{[s]}\}$ be the set of the diagonal elements and the links belonging to $G^{[s]}$ and define the set of *free elements* of $\boldsymbol{\Phi}^{[s]}$ as $\boldsymbol{\Phi}^{\nu(G^{[s]})} = \{\phi_{ij} \mid i, j \in \nu(G^{[s]})\}$. The remaining entries, that we simply refer to as non-free elements, are uniquely determined through the completion operation [1, Prop. 2] as a function of the free elements.

Suppose the proposed graph G' is obtained by adding edge (l, m) to the multigraph representation of $G^{[s]}$. The set of links that are changing in $\mathcal{G}$ is $L = \{(i, j) \mid i, j \in V, i < j, (i, j) \in E', (i, j) \notin E^{[s]}\}$. Its cardinality $l = |L|$ is arbitrary and, in general, different from one. We call $V(L) = B_l \cup B_m$ the set of the vertices involved in the change. Note that $\nu(G') = \nu(G^{[s]}) \cup L$. Our solution to define the new free elements is to maintain the same value for all the ones that are not involved in the change and to set the new ones by perturbing the current, non free elements, independently and all with the same variance σ_g^2. Namely, draw $\eta_h \sim^{ind} N(\phi_h^{[s]}, \sigma_g^2)$ and set $\phi'_h = \eta_h$ for each $h \in L$. Then, it is enough to derive all non free elements of Φ' though completion operation and finally to set $\boldsymbol{K}' = (\boldsymbol{\Phi}')^T \boldsymbol{\Phi}$. Note that, by doing so, we are generating a random variable $\boldsymbol{\eta}$ of length l that matches the dimension gap between $\boldsymbol{K}$ and $\boldsymbol{K}'$. As usual, the dimension decreasing case is deterministically defined in terms of the dimension increasing one.

4 Simulation Study

We compare our performances to the *Birth and Death* approach (BDMCMC for short) proposed by Mohammadi and Wit [14] and available in the R package `BDgraph` [16].

All final estimates, both from BDRJ and BDMCMC outputs, were obtained by controlling the *Bayesian False Discovery Rate*, as presented in [18] and [3]. Performances are assessed in terms of the standardized Structural Hamming Distance (Std-SHD, [23]) and the F_1-score [2, 19]. The first one prefers lower values, the second one higher values. Following the same approach of [14, 24], precision matrix estimation is measured using one half of the Stein loss score (SL) [8] which is equal to the Kullback-Leibler divergence [10] between $N_p(\boldsymbol{0}, \boldsymbol{K}_{true}^{-1})$ and $N_p(\boldsymbol{0}, \widehat{\boldsymbol{K}}^{-1})$.

In the first experiment, we set $p = 40$, $n = 500$ and $M = p/2$ groups of equal size, which leads to off-diagonal blocks of size 2×2. The true underlying graph is itself a block structured graph (see Fig. 2), while the true precision matrix was sampled by drawing from a G-Wishart $\left(3, \boldsymbol{I}_p\right)$. σ_g^2 was set equal to 0.5 after a little tuning phase. 400,000 iterations were run plus 100,000 extra iterations as burn-in period that were discarded. A simple visual inspection of Fig. 2 suggests that BDRJ is more precise than BDMCMC. The number of misclassified edges is rather low, Std-SHD $= 0.0243$, and it is well balanced between false positiveness (10) and negativeness (9). Many true discoveries are achieved and indeed it has F_1-score $= 0.954$.

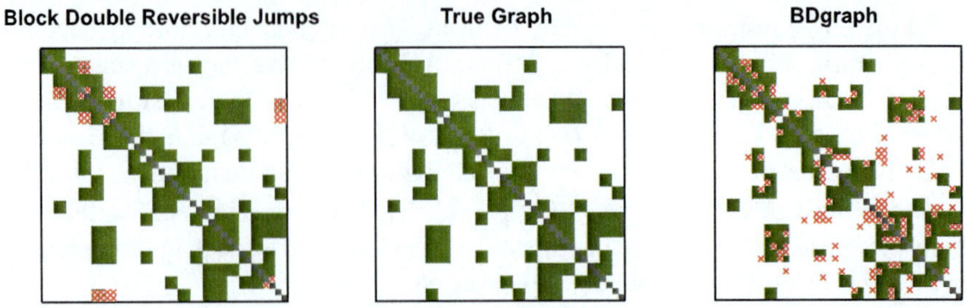

Fig. 2 The adjacency matrices of true underlying graph (middle panel), the BDRJ one (leftmost panel) and the one obtained using `BDgraph` (rightmost panel). Squares represent the included links, crosses stand for edges that are wrongly classified

BDMCMC does not recognize the block structure of the true graph, actually it does not even look for such a structure because the prior information can not be included in the model. It estimates the probability of inclusion of every possible link independently from the others. This entails more errors in the final estimate as well as a less informative structure of the graph. It would be hard to explain why there are missing edges within some structures that are clearly blocked ones. We repeated the same experiments for 18 different dataset: the true underlying graphs were randomly generated by sampling from (4) with different sparsity indices θ uniformly distributed in $[0.2, 0.6]$. The mean values, along with the standard deviations, for the F_1-scores are 0.845(0.13) and 0.80(0.03) and for the Std-SHD we have 0.053(0.04) and 0.060(0.02), respectively for BDRJ and BDMCMC. We see that BDRJ is more unstable with respect to BDMCMC. This is probably due to the fact that we used the same σ_g^2 for all dataset, without tuning it every time. However both indices prefer BDRJ.

The second experiment is inspired by a simulation study presented in [11] that aims to learn a graph and precision matrix under a noisy setting. The true underlying graph G is displayed in Fig. 3. We sample $\boldsymbol{K}_{\text{true}}|G \sim$ G-Wishart$(3, \boldsymbol{I}_p)$ and set $\boldsymbol{K}_{\text{noisy}}$ to be a random perturbation of $\boldsymbol{K}_{\text{true}}$: every possible value is perturbed, with probability s, by adding a random noise $0.1u$. Here $u \sim \text{Unif}\,(-k^*, k^*)$, where $k^* = \max_{i<j}|k_{ij}^{\text{true}}|$. Finally, data are generated from $N_p\left(\boldsymbol{0}, \boldsymbol{K}_{\text{noisy}}^{-1}\right)$. To investigate the behaviour under different volumes of noise, $s = 0.10, 0.20, 0.25$, we repeat each experiment 15 times. Results are reported in Table 1.

BDRJ outperforms BDMCMC on every dataset and with respect to all indices we considered. Its robustness is due to the fact that to conclude that a whole block has to be inserted in the final graph a single, isolated link is not enough. Those isolated values are not compatible with the block structured graph that BDRJ is looking for, therefore they are rightly ignored. On the other hand, BDMCMC does not look for any particular structure, hence it does not recognize the perturbed values as noise.

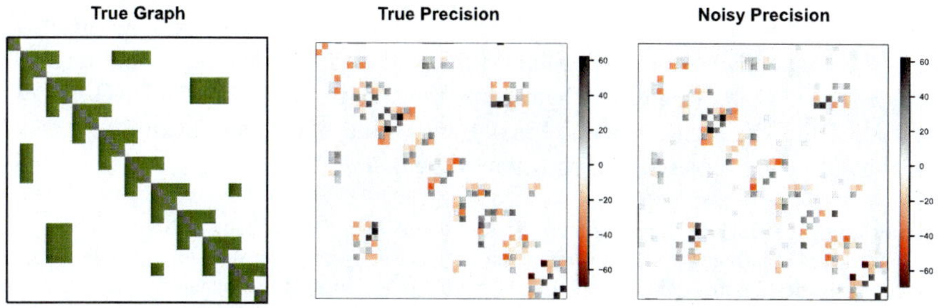

Fig. 3 The true underlying graph (leftmost panel) used to generate the true precision matrix $\boldsymbol{K}_{\text{true}}$ (middle panel). The rightmost panel is $\boldsymbol{K}_{\text{noisy}}$ (obtained with $s = 0.25$). For plotting purposes, we removed the diagonal in both precision matrices

Table 1 F_1-score, Std-SHD and Stein loss SL under different noise levels s

		s = 0.1	s = 0.2	s = 0.25
F1-score	BDRJ	**0.75 (0.035)**	**0.70 (0.043)**	**0.68 (0.043)**
	BDMCMC	0.44 (0.022)	0.405 (0.028)	0.066 (0.042)
SHD	BDRJ	**0.039 (0.004)**	**0.047 (0.043)**	**0.049 (0.005)**
	BDMCMC	0.063 (0.002)	0.36 (0.042)	0.071 (0.003)
SL	BDRJ	**0.24 (0.030)**	**0.32 (0.041)**	**0.37 (0.060)**
	BDMCMC	1.00 (0.032)	1.05 (0.036)	1.08 (0.059)

Values in bold are the ones preferred by the corresponding index

5 Discussion

In the setting of graphical models, this work proposed a new class of priors, called *block graph priors*. They allow to include in the model the prior knowledge available about the partition of the nodes. We also introduced a new sampling strategy that leverage these priors to look only for a block structured graph, whose block, if included, have to be complete. In some applications, as the number of variables grows, the importance of each possible dependence loses of interest as it is more natural, and more interpretable, to understand the general structure of dependencies. This is the case of genomics applications as genes may be grouped in pathways, therefore a block structured graph is expected and more interpretable. Another example is market basket analysis which aims to find patterns of association between retailed items so that they can be bundled together to the end of delivering an appealing offer. Finally, we compared our model, on synthetic data, with BDMCMC. In both experiments, BDRJ estimates are better in terms of Std-SHD, F_1-score and SL.

As future developments, we aim to further develop the BDRJ technique, expand the simulation study by investigating the behaviour of BDRJ when the underlying graph has incomplete blocks and to assess its performances in real world applications. Moreover, experiments are performed using groups of only two nodes. Larger groups imply larger jumps of the chain in the state space and therefore they are less likely to be accepted. We aim to better investigate the behaviour of our methodology in such cases. We would also like to understand if the proposed methodology could be also extended to Gaussian structured chain graph models for modelling the DAG model induced by the chain components. Finally, we would like to add flexibility to the model by allowing for a random partition of the nodes.

Acknowledgements The author is grateful to Raffaele Argiento (Universitá di Bergamo), Lucia Paci and Alessia Pini (both affiliated to Universitá Cattolica del Sacro Cuore) for the valuable advice received in the development of the work and during the preparation of the manuscript as well.

References

1. Atay-Kayis, A., Massam, H.: A Monte Carlo method for computing the marginal likelihood in nondecomposable Gaussian graphical models. Biometrika **92**, 317–335 (2005)
2. Baldi, P., Brunak, S., Chauvin, Y., Andersen, C.A., Nielsen, H.: Assessing the accuracy of prediction algorithms for classification: an overview. Bioinformatics **16**(5), 412–424 (2000)
3. Codazzi, L., Colombi, A., Gianella, M., Argiento, R., Paci, L., Pini, A.: Gaussian graphical modeling for spectrometric data analysis. Comput. Stat. Data Anal. (2022)
4. Cremaschi, A., Argiento, R., De Iorio, M., Shirong, C., Chong, Y.S., Meaney, M.J., Kee, M.Z.: Seemingly unrelated multi-state processes: a Bayesian semiparametric approach. arXiv preprint arXiv:2106.03072 (2021)
5. Dobra, A., Hans, C., Jones, B., Nevins, J.R., Yao, G., West, M.: Sparse graphical models for exploring gene expression data. J. Multivariate Anal. **90**(1), 196–212 (2004)
6. Giudici, P., Castelo, R.: Improving Markov chain Monte Carlo model search for data mining. Mach. Learn. **50**(1–2), 127–158 (2003)
7. Giudici, P., Green, P.: Decomposable graphical Gaussian model determination. Biometrika **86**(4), 785–801 (1999)
8. James, W., Stein, C.: Estimation with quadratic loss. In: Proceedings of the Fourth Berkeley Symposium on Mathematical Statistics and Probability, vol. 1, pp. 361–379 (1961)
9. Jones, B., Carvalho, C., Dobra, A., Hans, C., Carter, C., West, M.: Experiments in stochastic computation for high-dimensional graphical models. Stat. Sci. **20**, 388–400 (2005)
10. Kullback, S., Leibler, R.A.: On information and sufficiency. Ann. Math. Stat. **22**(1), 79–86 (1951)
11. Kumar, S., Ying, J., de Miranda Cardoso, J.V., Palomar, D.P.: A unified framework for structured graph learning via spectral constraints. J. Mach. Learn. Res. **21**(22), 1–60 (2020)
12. Lauritzen, S.L.: Graphical Models. Oxford University Press, Oxford (1996)
13. Lenkoski, A.: A direct sampler for G-Wishart variates. Statistics **2**(1), 119–128 (2013)
14. Mohammadi, A., Wit, E.C.: Bayesian structure learning in sparse Gaussian graphical models. Bayesian Anal. **10**(1), 109–138 (2015)
15. Mohammadi, R., Massam, H., Letac, G.: Accelerating Bayesian structure learning in sparse gaussian graphical models. J. Am. Stat. Assoc. **0**(0), 1–14 (2021)
16. Mohammadi, R., Wit, E.C.: BDgraph: an R package for Bayesian structure learning in graphical models. J. Stat. Software **89**(3), 1–30 (2019). https://doi.org/10.18637/jss.v089.i03
17. Osborne, N., Peterson, C., Vannucci, M.: Latent network estimation and variable selection for compositional data via variational EM. J. Comput. Graph. Stat. **31**(1), 1–22 (2021)
18. Peterson, C., Stingo, F.C., Vannucci, M.: Bayesian inference of multiple Gaussian graphical models. J. Am. Stat. Assoc. **110**(509), 159–174 (2015)
19. Powers, D.M.W.: Evaluation: from precision, recall and F-measure to ROC, informedness, markedness and correlation. J. Mach. Learn. Technol. **2**, 2229–3981 (2011)

20. Roverato, A.: Hyper inverse Wishart distribution for non-decomposable graphs and its application to Bayesian inference for Gaussian graphical models. Scand. J. Stat. **29**(3), 391–411 (2002)
21. Scott, J., Carvalho, C.: Feature-inclusion stochastic search for Gaussian graphical models. J. Comput. Graph. Statist. **17**(4), 790–808 (2008)
22. Scutari, M.: On the prior and posterior distributions used in graphical modelling. Bayesian Anal. **8**(3), 505–532 (2013)
23. Tsamardinos, I., Brown, L.E., Aliferis, C.F., Moore, A.W.: The max-min hill-climbing Bayesian network structure learning algorithm. Mach. Learn. **65**(1), 31–78 (2006)
24. Wang, H.: Sparse seemingly unrelated regression modelling: applications in finance and econometrics. Comput. Stat. Data Anal. **54**(11), 2866–2877 (2010)
25. Wang, H.: Scaling it up: stochastic search structure learning in graphical models. Bayesian Anal. **10**(2), 351–377 (2015)
26. Wang, H., Li, S.Z.: Efficient Gaussian graphical model determination under G-Wishart prior distributions. Electron. J. Stat. **6**, 168–198 (2012)
27. Xia, Y., Cai, T., Cai, T.T.: Multiple testing of submatrices of a precision matrix with applications to identification of between pathway interactions. J. Am. Stat. Assoc. **113**(521), 328–339 (2018)

A Bayesian Joint Spatio-temporal Model for Multiple Mosquito-Borne Diseases

Jessica Pavani and Paula Moraga

Abstract Many infectious diseases studied in the epidemiological context are caused by insects, mainly mosquitoes. These infections are known as arboviruses because they need vectors to be transmitted. Some of them may be related to each other since a same mosquito species can transmit different diseases. This study aims to describe geographic and temporal patterns of two mosquito-borne diseases, dengue and chikungunya, and their possible risk factors in the Brazilian state of Ceará in 2017. To pursue this, we consider a Bayesian hierarchical spatio-temporal model for the joint analysis of both arboviruses. This specification also uses a Zero-Inflated Poisson (ZIP) model to overcome the high proportion of zeros. Moreover, it includes covariates as well as disease-specific and shared spatial and temporal effects, which are estimated and mapped to identify similarities among diseases. Our findings help understand geographic and temporal disease patterns, and to identify high risk areas and potential risk factors, and can inform the development and implementation of strategies for disease prevention and control.

Keywords Bayesian model · Chikungunya fever · Dengue fever · INLA · Multivariate disease mapping · Spatial modeling

1 Introduction

In an epidemiological investigation, the understanding of the connection between disease occurrence and its geographic and temporal trends can help decision-makers to develop strategies for disease prevention and control [10]. In case of mosquito-

J. Pavani (✉)
Department of Statistics, Pontificia Universidad Católica de Chile, Santiago, Chile
e-mail: jlpavani@mat.uc.cl

P. Moraga
Computer, Electrical and Mathematical Sciences and Engineering Division, King Abdullah University of Science and Technology, Thuwal, Saudi Arabia
e-mail: paula.moraga@kaust.edu.sa

R. Argiento et al. (eds.), *New Frontiers in Bayesian Statistics*, Springer Proceedings in Mathematics & Statistics 405, https://doi.org/10.1007/978-3-031-16427-9_7

borne diseases, many spread characteristics are indistinguishable. Geographic and temporal patterns are also similar. This occurs because some diseases are transmitted by the same mosquito species, whose breeding and development are usually influenced by climatic factors such as rainfall, humidity, and temperature [5]. Thus, it is reasonable to think that both the geographical and the temporal patterns of these diseases could show common features, and a potential autocorrelation between them could exist.

Although the evolution of mosquito-borne diseases have been widely explored, most of the studies focus on modeling arboviruses separately [2, 12]. Despite that, some joint modeling approaches can be found in the literature. Freitas et al. [6] and Kazazian et al. [8] investigated three mosquito-borne diseases outbreaks (dengue, Zika, and chikungunya) in different Brazilian regions looking for simultaneous clustering patterns. Carvalho et al. [3], on the other hand, analyzed the association among the Zika epidemic and past dengue outbreaks in the same Brazilian region.

In this study, we jointly analyze dengue and chikungunya, two of the mosquito-borne diseases that co-circulate in Brazil. Our goal is to identify geographic and temporal patterns as well as potential risk factors. We consider the number of weekly cases in 2017 in each municipality of the Brazilian state of Ceará, one of the states with higher prevalence of both arboviruses. We use a Bayesian joint model that includes spatio-temporal covariates, known to affect disease transmission, as well as random effects to model residual variation, and considers the potential autocorrelation between the two diseases. The model also uses a Zero-Inflated Poisson (ZIP) formulation to overcome the high proportion of zeros.

The rest of this manuscript is organized as follows. Section 2 presents the formulation of the Bayesian spatio-temporal approach. In Sect. 3, we provide a description of the dataset that motivates our analysis and more details about the study area and the covariates. Section 4 is dedicated to the results, and it includes details about the risk factors, spatial and temporal effects, as well as the relative risks estimates for each municipality in specific epidemiological weeks. Finally, in Sect. 5 we present a discussion including limitations of our modeling approach that motivate future work.

2 Spatio-temporal Modeling

Different ways of inducing spatial and temporal correlation may be found in the literature. They usually consist of defining a prior distribution with some dependence structure. Conditional autoregressive (CAR) prior is commonly used in spatial studies, while temporal analysis are typically based on autoregressive or random walk structures. Such strategies may be extended to deal with multivariate dataset. In the Bayesian disease mapping context, Moraga and Lawson [11], for instance, reviewed two approaches used to induce the spatial dependence between regions, and also the dependence among multiple diseases. On the other hand, Gómez-Rubio et al. [7] proposed a spatio-temporal specification where both spatial and temporal random effects are built considering disease-specific and shared components.

This study aims to jointly model multiple diseases based on a specific and shared spatial and temporal effects approach. Furthermore, in the context of arboviruses, it is important to consider that cases of the disease are sparse at certain regions and/or periods of the year. Therefore, in order to overcome the high proportion of zeros, our specification uses a Zero-Inflated Poisson (ZIP) regression model. This modeling combines the proportion of zeros with Poisson (Poi) distribution, so that the probability function can be formulated as follows:

$$P[O_{i,t} \mid p_{i,t}, \theta_{i,t}, E_{i,t}] = p_{i,t}\mathbb{1}_{[O_{i,t}=0]} + (1 - p_{i,t})Poi(\theta_{i,t}E_{i,t}),$$

where $p_{i,t}$ is the probability of extra zeros, $O_{i,t}$ is the observed number of cases, $E_{i,t}$ is the expected number of cases computed via internal standardization [1, Chapter 11], and $\theta_{i,t}$ is the relative risk, for area i and time t.

Note that the interest here is to model two latent fields, $p_{i,t}$ and $\theta_{i,t}$, using the canonical link function $\text{logit}(p_{i,t})$ and $\log(\theta_{i,t})$, where both can include covariates and random effects. However, we incorporate spatio-temporal covariates only to the Poisson component (see Sect. 3 for the dataset description), so the final model is formulated as follows:

$$O_{i,t}^{(d)} \mid p_{i,t}, E_{i,t}^{(d)}, \theta_{i,t}^{(d)} \sim \text{ZIP}(p_{i,t}, E_{i,t}^{(d)}\theta_{i,t}^{(d)}) \quad (1)$$

$$\log(\theta_{i,t}^{(d)}) = \alpha^{(d)} + \boldsymbol{\beta}^{\top} X_{i,t} + \Phi_i^{(d)} + \Psi_t^{(d)} \,. \quad (2)$$

In this notation, (d) represents the disease, $\alpha^{(d)}$ are disease-specific intercepts and $\boldsymbol{\beta}$ are regression coefficients related to the spatio-temporal covariates, $X_{i,t}$. Finally, $\Phi_i^{(d)}$ and $\Psi_t^{(d)}$ are spatial and temporal effects also considered for area i, time t, and disease d. These effects are composed of disease-specific and shared patterns, as follows:

$$\Phi_i^{(d)} = u_i^{(d)} + \delta_S^{(d)} U_i \qquad u_i^{(d)} \sim \text{CAR}(W, \tau_S^{(d)}) \qquad U \sim \text{CAR}(W, \tau_{0S}) \quad (3)$$

$$\Psi_t^{(d)} = v_t^{(d)} + \delta_T^{(d)} V_t \qquad v_t^{(d)} \sim \text{CAR}(Q, \tau_T^{(d)}) \qquad V \sim \text{CAR}(Q, \tau_{0T}) \,. \quad (4)$$

In this case, U_i and V_t represent spatial and temporal effects shared by the diseases, while $u_i^{(d)}$ and $v_t^{(d)}$ are spatial and temporal effects specific for each disease. $\delta_S^{(d)}$ and $\delta_T^{(d)}$ work as weights to control the shared effects on the relative risk. The model formulation is completed by assuming a CAR specification to both disease-specific and shared effects, where W and Q are spatial and temporal adjacency matrices, and $\tau_S^{(d)}$, $\tau_T^{(d)}$, τ_{0S}, and τ_{0T} denote the precision of each effect.

For Bayesian inference, although Markov chain Monte Carlo (MCMC) methods are extensively used they present some limitations, mainly related to computational time and convergence implications. In this case, the complexity of the spatio-temporal structure combined with a large dataset could lead to several days of computing time to perform Bayesian inference via MCMC. To overcome this issue, the integrated nested Laplace approximation was considered [13].

Regarding the prior distributions, we follow the suggestions by Gómez-Rubio et al. [7] and assign flat priors to the disease-specifics intercepts and to the precision components of spatial and temporal effects. For the spatial and temporal weights, log-Normal priors are defined, which restricts them to be positive.

3 Motivational Data

As motivation to this study, we consider two mosquito-borne diseases that co-circulate in Brazil, namely dengue and chikungunya. The dataset consists of the total number of cases of each disease reported per municipality in Ceará and epidemiological week. Such information has been collected by Infodengue, a system that computes the clinically confirmed cases that are reported by medical professionals through official channels. See Codeco et al. [4] for more details about the system and data collection.

3.1 Study Area

Located in South America, Brazil is politically divided into 27 administrative states, being Ceará the eighth most populous. According to the Brazilian Institute of Geography and Statistics, in 2020 the state reached an estimated population of 9.2 million. Administratively, the state is divided into 184 municipalities, mostly with population under 50 thousands people. These 184 municipalities are the areal units contemplated in this study.

In regard to mosquito-borne diseases, Ceará presents favorable environmental and socio-demographic conditions for mosquitoes breeding [12]. Indeed, the state has been faced to recurrent outbreaks of different arboviruses. In this study, we consider two of the most prevalent, dengue and chikungunya. These diseases are transmitted by the same mosquito species (Aedes aegypti) and co-circulate over the Ceará state since 2016. To complete the spatio-temporal design, we considered the year of 2017, which is divided in 52 epidemiological weeks and are the temporal units in this study.

3.2 Risk Factors

It is known that environmental factors play an important role in the spread of arboviruses, since breeding and development of mosquitoes are influenced by climatic factors. In this study, we consider two climatic factors as covariates, the minimum temperature (Celsius degree) and the maximum humidity (percentage). This information was considered per municipality and week, and was collected by the nearest airports [4].

During the study period, it was observed minimum temperature between 19°C in winter and 26°C in summer. At this point, it could be important to highlight that in Brazil the winter occurs between June, 21st and September 21st while the summer occurs between December 21st and March 21st. Similarly, the maximum humidity ranged between 79 and 97%, reaching higher values between March and May in regions with tropical wet climate. As noted, there was little variation in both temperature and humidity. This is because the state of Ceará has a predominant semi-arid climate. Only part of the coast and areas with the highest topographical elevation present a tropical wet climate. This condition guarantees warm temperatures, fairly constant throughout the year and little variation in humidity, since the semi-arid climate is known for periodic droughts and low rainfall.

4 Results

The effect of predictors on the risk of dengue and chikungunya as well as disease-specific intercepts are summarized in Table 1. The posterior mean of the effect of the predictors indicated that temperature is positively related to the diseases, while the humidity is negatively related to them. Regarding the disease-specific intercepts, negative values were found to both infections. However, while the posterior mean estimates for the dengue-intercept is approximately −1.9, for the chikungunya-intercept is approximately −3.1, showing difference between diseases. With respect to the ZIP parameter, its posterior mean is 0.264 with 2.5 and 97.5 quantiles of 0.250 and 0.277, respectively.

4.1 Spatial and Temporal Effects

In this section, we describe the patterns of spatial and temporal effects. Posterior means of the total spatial effect, as presented by Eq. (3), can be seen in Fig. 1. We observe dengue and chikungunya show different spatial patterns, especially in the central area of the Ceará. However, the highest risk areas for both diseases tend to

Table 1 Summary statistics of the disease-specific and predictors—mean, standard deviation (SD), 2.5 , 50, and 97.5 quantiles

Predictors	Mean	SD	2.5% q.	50% q.	97.5% q.
Intercept for dengue	−1.946	0.071	−2.086	−1.946	−1.807
Intercept for chikungunya	−3.127	0.077	−3.280	−3.127	−2.977
Temperature	0.003	0.002	−0.001	0.003	0.006
Humidity	−0.002	0.000	−0.003	−0.002	−0.001

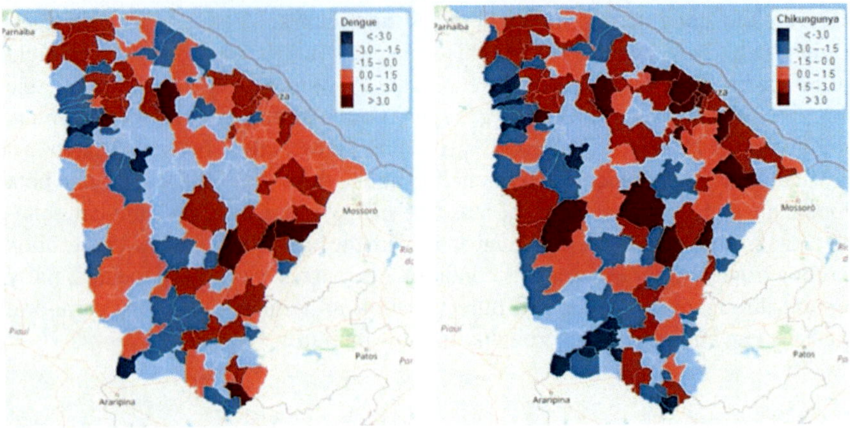

Fig. 1 Posterior mean of the spatial effect Φ_i of dengue (left) and chikungunya (right).

Table 2 Summary statistics of the weights for shared spatial and temporal effects - mean, standard deviation (SD), 2.5 , 50, and 97.5 quantiles. (1) indicates dengue and (2) chikungunya.

Parameter	Mean	SD	2.5% q.	50% q.	97.5% q.
$\delta_S^{(1)}$	1.308	0.122	1.064	1.311	1.540
$\delta_S^{(2)}$	1.675	0.103	1.478	1.673	1.880
$\delta_T^{(1)}$	0.025	0.018	0.003	0.021	7.100
$\delta_T^{(2)}$	0.504	1.748	0.009	0.146	3.180

concentrate in the northeast of the state, mainly on the coastal region. The spatial weights, available in Table 2, are similar for the diseases ($\delta_S^{(1)} = 1.31$ and $\delta_S^{(2)} = 1.67$, where (1) indicates dengue and (2) chikungunya). Hence, the spatial pattern of each disease is similar to their shared pattern. Overall, weights greater than one indicate a high dependence on the shared spatial effect.

Regarding the total temporal effect, represented as Ψ_t on Eq. (4), it is also possible to notice a different pattern between the diseases, Fig. 2. Dengue pattern indicates a decrease in risk over time. On the other hand, the temporal effect of chikungunya has an increase behavior until reach its peak on 20th week, and then it starts falling. Differently from the spatial weights, the temporal weights have values close to zero, indicating a low dependence on the shared effect ($\delta_T = 0.02$ for dengue and $\delta_T = 0.50$ for chikungunya, Table 2).

4.2 Relative Risk

Throughout the year, 66,083 cases of dengue and 98,933 cases of chikungunya were reported in the state of Ceará. These cases were most concentrated between 12th

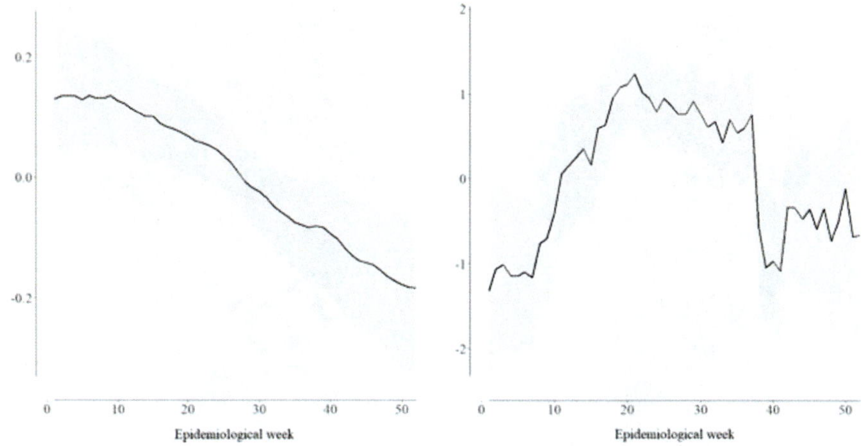

Fig. 2 Posterior mean of the temporal effect Ψ_t of dengue (left) and chikungunya (right)

and 23rd epidemiological week, i.e., months of April and May, approximately. The highest number of dengue cases occurred in the 15th week, in which 4939 cases were reported. Regarding chikungunya, the peak was in the 19th week with 8421 reported cases. In both situations, the capital Fortaleza was responsible for more than 70% of the total cases.

For practical reasons, estimates of relative risk for each municipality were mapped only for the weeks that presented peak of cases for each disease. Thus, Fig. 3 corresponds to the 15th week, while Fig. 4 is related to the 19th week. Most of the municipalities have a relative risk less than one for both diseases. As expected, there

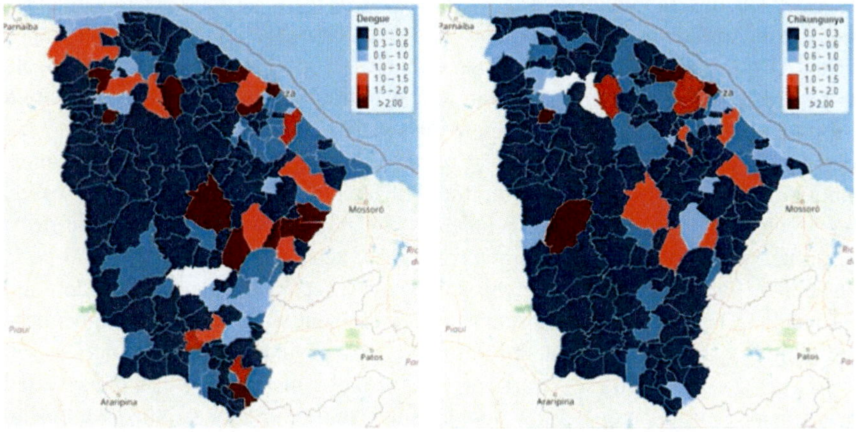

Fig. 3 Posterior mean relative risk estimates of dengue (left) and chikungunya (right) in the Brazilian state of Ceará—15th epidemiological week

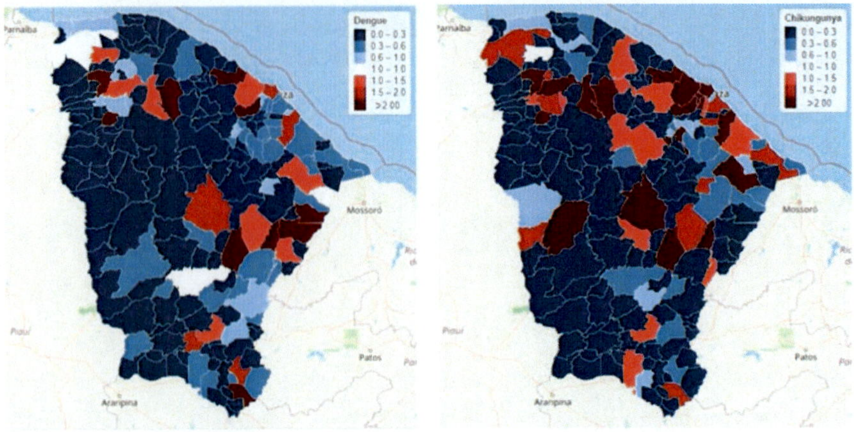

Fig. 4 Posterior mean relative risk estimates of dengue (left) and chikungunya (right) in the Brazilian state of Ceará—19th epidemiological week

are more high risk areas for dengue in week 15th and more high risk areas for chikungunya in week 19th. Of the 184 municipalities, only 11 have an over one relative risk for both diseases in the 15th week. This number slightly increases in the 19th week, with 16 municipalities.

5 Discussion and Future Work

This study shows an analysis of how to jointly model mosquito-borne diseases using Bayesian spatio-temporal modeling. We have chosen to use an approach that includes both spatial and temporal effects, considering disease-specific and shared components. As covariates, we have used two climate factors that play an important role in the spread of arboviruses, temperature and humidity. The model has been fitted using the Bayesian computational approach INLA.

Overall, we have obtained interesting results. Different disease-specific intercepts were found, which suggests an initial difference between the risk of diseases. Temperature and humidity were included as covariates in the model. Data for these covariates were collected from airports stations, and since the number of airports available in the region is not large, this could have affected the quality of the data and our results. Therefore, in future studies, more efficient ways to collect temperature and humidity should be considered. Moreover, the inclusion of interactions as well as non-linear effects of these covariates could also be considered. This will be treated as extensions for this model and would help us to better understand the relationship between mosquito-borne diseases and potential risk factors.

Regarding spatial and temporal patterns, we have noticed different behaviors between diseases. When considering more closely, the weeks when the dengue and chikungunya peaks occurred, only 6% and 9% of the 184 municipalities shown relative risk greater than one, respectively. These results indicate a possible competition between viruses since epidemiological knowledge indicates the diseases tend not to occur at the same time and space. However, these results should be cautiously considered given that the syndromes of the two diseases are similar, which could lead to misdiagnosis. Furthermore, a longer time period should be considered to verify whether this pattern is maintained. The inclusion of space-time interactions is also a possible extension to this model [9].

To conclude, future work will be devoted to implement the extensions mentioned and address some limitations in data and modeling. The joint spatio-temporal model developed allows us to better understand the geographic and temporal spread of diseases and can help policymakers in the development of strategies for disease prevention and control.

References

1. Banerjee, S., Carlin, B.P., Gelfand.: Hierarchical Modeling and Analysis for Spatial Data. Taylor & Francis Inc., A. E. (2011)
2. Carmo, R.F., Júnior, J.V.J.S., Pastor, A.F., Souza, C.D.F.: Spatiotemporal dynamics, risk areas and social determinants of dengue in Northeastern Brazil, 2014–2017: an ecological study. Infect. Dis. Poverty **9**(1), 153 (2020)
3. Carvalho, M.S., Freitas, L.P., Cruz, O.G., Brasil, P., Bastos, L.S.: Association of past dengue fever epidemics with the risk of Zika microcephaly at the population level in Brazil. Sci. Rep. **10**(1), 1752 (2020)
4. Codeco, C., Coelho, F., Cruz, O., Oliveira, S., Castro, T., Bastos, L.: Infodengue: a nowcasting system for the surveillance of arboviruses in Brazil. Rev. Epidemiol. Sante Publique **66**(5), S386 (2018)
5. Franklinos, L., Jones, K., Redding, D., Abubakar, I.: The effect of global change on mosquito-borne disease. Lancet. Infect. Dis. **19**(9), e302–e312 (2019)
6. Freitas, L.P., Cruz, O.G., Lowe, R., Carvalho, M.S.: Space-time dynamics of a triple epidemic: dengue, chikungunya and Zika clusters in the city of Rio de Janeiro. Proc. Biol. Sci. **286**(1912), 20191867 (2019)
7. Gómez-Rubio, V., Palmí-Perales, F., López-Abente, G., Ramis-Prieto, R., Fernández-Navarro, P.: Bayesian joint spatio-temporal analysis of multiple diseases. SORT **43**(1), 51–74 (2019)
8. Kazazian, L., Neto, A.S.L., Sousa, G.S., Nascimento, O.J., Castro, M.C.: Spatiotemporal transmission dynamics of co-circulating dengue, Zika, and chikungunya viruses in Fortaleza, Brazil: 2011–2017. PLoS Negl. Trop. Dis. **14**(10), e0008760 (2020)
9. Knorr-Held, L.: Bayesian modelling of inseparable space-time variation in disease risk. Stat. Med. **19**(17–18), 2555–2567 (2000)
10. Moraga, P.: Geospatial Health Data: Modeling and Visualization with R-INLA and Shiny. Chapman & Hall/CRC (2019)
11. Moraga, P., Lawson, A.B.: Gaussian component mixtures and CAR models in Bayesian disease mapping. Comput. Stat. Data Anal. **56**(6), 1417–1433 (2012)
12. Rodrigues, N.C.P., Lino, V.T.S., Daumas, R.P., Noronha-Andrade, M.K., O'Dwyer, G., Monteiro, D.L.M., Gerardi, A., Fernandes, G.H.B.V., Ramos, J.A.S., Ferreira, C.E.G., Costa-Leite, I.: Temporal and spatial evolution of dengue incidence in Brazil, 2001–2012. PLOS ONE **11**(11), e0165945 (2016)
13. Rue, H., Martino, S., Chopin, N.: Approximate Bayesian inference for latent Gaussian models by using integrated nested Laplace approximations. J. R. Statist. Soc. B **71**(2), 319–392 (2009)

A Bayesian Nonparametric Test for Cross-Group Differences Relative to a Control

Iván Gutiérrez, Luis Gutiérrez, and Danilo Alvares

Abstract We propose a new Bayesian nonparametric multivariate testing procedure for comparing several treatments against a control. The test is based on a general model where the distribution of each treatment group can be identical to (or different from) the control group distribution, depending on the value of a latent binary vector. This vector is endowed with a spike-and-slab prior distribution carefully chosen to ensure a multiplicity correction. Group distributions are modeled in a flexible way using a dependent Dirichlet process. Monte Carlo experiments suggest that our proposal performs better than state-of-the-art frequentist alternatives for small sample sizes.

Keywords Dependent Dirichlet process · MANOVA · Multiple testing · Spike-and-slab prior

1 Introduction

Comparing the underlying multivariate distributions of $J + 1$ groups is a common problem in applied statistics. Typically, we want to detect differences among any pair of distributions (tests that tackle this problem are called *k-sample tests*). However, in some situations, we only want to compare J of these groups (e.g., the treatment groups) against the remaining one (e.g., the control group). Hypothesis tests that handle this problem are known as *k versus 1-sample tests*.

I. Gutiérrez (✉) · L. Gutiérrez · D. Alvares
Department of Statistics,Pontificia Universidad Católica de Chile,4860 Vicuña Mackenna, Santiago, Chile
e-mail: isgutierrez@mat.uc.cl

L. Gutiérrez
e-mail: llgutier@mat.uc.cl

D. Alvares
e-mail: dalvares@mat.uc.cl

R. Argiento et al. (eds.), *New Frontiers in Bayesian Statistics*, Springer Proceedings in Mathematics & Statistics 405, https://doi.org/10.1007/978-3-031-16427-9_8

Historically, the k versus 1-sample test has been based on the multivariate analysis of variance (MANOVA) [3]. Even today, MANOVA remains popular because is easy to understand, and is currently implemented in all major statistical software (e.g., R, SAS, SPSS, Stata and so on). However, MANOVA can only detect differences across the groups means.

The desire to detect more types of differences has motivated several new tests in recent years. From the frequentist perspective, there are flexible univariate 2-sample tests [31], multivariate 2-sample tests [4, 5, 11, 25], and multivariate k-sample tests [22]. However, in order to adapt these procedures for k versus 1-sample tests, we need to use post-hoc adjustments (e.g., a multiplicity correction); otherwise, part of the power would be wasted trying to detect differences across the treatment groups.

From the Bayesian perspective, there exist flexible Bayesian non-parametric models specifically designed to borrow strength across several multivariate distributions [23, 28, 30], but all these models treat the $J + 1$ groups symmetrically and do not include a formal testing procedure. In addition, there exist flexible 2-sample tests [16, 19] and k-sample tests [6, 7], all of them based on Polya trees [1] and Dirichlet processes (DP) [10]. More recently, Gutiérrez et al. [14] developed a univariate k versus 1-sample test based on DP with an absolutely continuous spike-and-slab (SS) prior [21]. However, the choice of an absolutely continuous SS prior implies that the groups distributions can never be identical but only *similar*. Moreover, the original idea cannot be directly generalized to the multivariate case because many important conjugacy properties would be lost. We still could adapt the idea, but the posterior inference would depend on a difficult-to-tune MCMC algorithm, making the testing procedure more fragile (from a computational point of view) and time consuming.

In this article, we present a flexible Bayesian nonparametric multivariate k versus 1-sample test that solves all the aforementioned problems. First, we replace the original SS prior with a Dirac SS prior, enabling *identical* groups' distributions under the spike. Second, we move the SS prior from the DP base distribution to a higher level of the model hierarchy, enabling the marginalization of several parameters and the use of more efficient samplers. Our proposal differs from Gutiérrez et al. [14] model in two critical ways. Firstly, the distribution of each treatment group becomes identical (instead of similar) to the distribution of the control group under the spike. Secondly, from a computational point of view, our model can accommodate multivariate responses without giving up its conjugacy properties, simplifying the posterior inference to a considerable extent, reducing the computational cost of the testing procedure. In addition, our model differs from all the other Bayesian k-sample tests in that, if the problem at hand is really a comparison of J treatments against a control, then the number of the admissible hypotheses is 2^J, which is much lower than the B_{J+1} admissible hypotheses in a k-sample test, where B_j is the jth Bell number. For example, if $J = 6$, $2^6 = 64 \ll 877 = B_7$. This reduction in the cardinality of the hypothesis space simplifies the posterior inference to a considerable extent.

The rest of the manuscript is organized as follows. In Sect. 2, we introduce our new BNP k versus 1-sample test. In Sect. 3, we explain how to learn the key parameters of this model. In Sect. 4, we compare our method with a state-of-the-art alternative thorough a simulation experiment. In Sect. 5, we conclude with a discussion and some direction for future work.

2 A BNP Model for Multivariate Comparisons

Let $\mathscr{D} = \{(y_i, x_i)\}_{i=1}^N$ be a sample, where $y_i \in \mathbb{R}^D$ is the D-variate response variable for the ith experimental unit and $x_i \in \mathscr{J} := \{0, \ldots, J\}$ represents the group indicator. There are two types of groups: one control group, labelled as *0*, and J treatment groups, labelled as $1, \ldots, J$. We want to know which treatment groups follow the same distribution as the control group. To do so, we consider the following model:

$$\begin{aligned} y_i \mid \{x_i = 0\}, Q_0 &\overset{iid}{\sim} Q_0, \\ y_i \mid \{x_i = j\}, Q_0, Q_j &\overset{iid}{\sim} \gamma_j Q_j + (1 - \gamma_j) Q_0, \end{aligned}$$

where $Q_0, \ldots, Q_J$ are random distributions and $\boldsymbol{\gamma} := (\gamma_1, \ldots, \gamma_J) \in \{0, 1\}^J$ is a vector of latent variables determining which groups follow the same distribution as the control group. So, each $\boldsymbol{\gamma}$ represents a different model or *hypothesis*.

Following a Bayesian framework, we can learn γ by computing

$$p(\boldsymbol{\gamma} \mid \boldsymbol{y}) = \left(\sum_{\boldsymbol{\beta} \in \{0,1\}^J} \frac{\pi_0(\boldsymbol{\beta})}{\pi_0(\boldsymbol{\gamma})} B_{\boldsymbol{\beta},\boldsymbol{\gamma}} \right)^{-1}, \tag{1}$$

where π_0 is the prior hypothesis distribution and $B_{\boldsymbol{\beta},\boldsymbol{\gamma}=L(\boldsymbol{y}|\boldsymbol{\beta})/L(\boldsymbol{y}|\boldsymbol{\gamma})}$ is the *Bayes factor* given two hypotheses $\boldsymbol{\beta}$ and $\boldsymbol{\gamma}$, with $L(\boldsymbol{y} \mid \boldsymbol{\gamma})$ being the marginal likelihood given $\boldsymbol{\gamma}$. This specification has two main advantages. Firstly, the Bayes factors control for model complexity in a natural, principled way [18]. Secondly, some judicious choices of π_0 apply a multiplicity correction [26, 32]. However, in practice, $L(\boldsymbol{y} \mid \boldsymbol{\gamma})$ can be difficult to compute, because it involves the marginalization of any additional variable. Hence, we need to find the balance between the flexibility of our model and its computational tractability.

2.1 A Prior for the Hypotheses

As in Womack et al. [32], we set a prior for $\pi_0(\cdot)$ such that, for any $\boldsymbol{\gamma}, \boldsymbol{\beta} \in \{0, 1\}^J$,

1. $\pi_0(\boldsymbol{\gamma}) = \zeta_0 \sum_{\boldsymbol{\beta}:\boldsymbol{\beta} \succ \boldsymbol{\gamma}} \pi_0(\boldsymbol{\beta})$,
2. $\|\boldsymbol{\gamma}\|_1 = \|\boldsymbol{\beta}\|_1 \Rightarrow \pi_0(\boldsymbol{\gamma}) = \pi_0(\boldsymbol{\beta})$,

for some fixed $\zeta_0 > 0$, where $\boldsymbol{\beta} \succ \boldsymbol{\gamma}$ means that $\boldsymbol{\beta} \neq \boldsymbol{\gamma}$, and $\beta_j \geq \gamma_j$ for all $j = 1, \ldots, J$. This distribution must exist, because each $\pi_0(\boldsymbol{\gamma})$ is a deterministic function

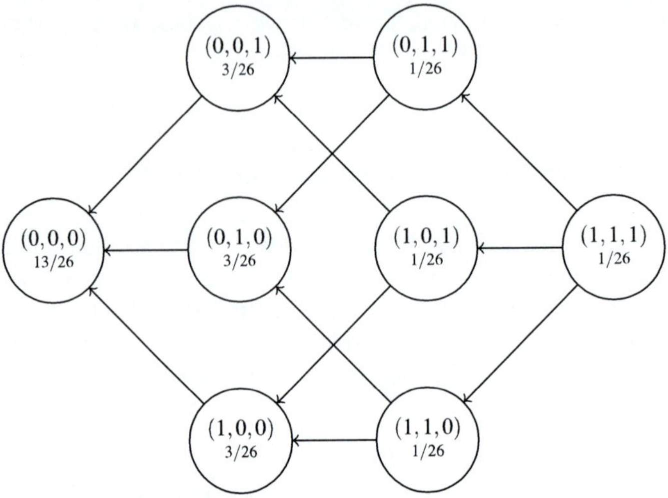

Fig. 1 Hasse diagram for the partially ordered set $(\{0, 1\}^3, \succ)$, alongside with the probability of each node (in our context, hypothesis) under the Womack distribution with $\zeta_0 = 1$ (under each vector). Given any hypotheses $\boldsymbol{\beta}$ and $\boldsymbol{\gamma}$, it can be proved that $\boldsymbol{\beta} \succ \boldsymbol{\gamma}$ if and only if there is a directed path from $\boldsymbol{\beta}$ to $\boldsymbol{\gamma}$. As some hypotheses are not connected by directed paths, they are incomparable

of $(\pi_0(\boldsymbol{\beta}) : \boldsymbol{\beta} \succ \boldsymbol{\gamma})$. Using this prior, the more complex hypotheses are penalized in a way that is easy to interpret. Figure 1 illustrates this phenomena:

Moreover, as all the $\boldsymbol{\gamma}$'s with the same 1-norm must have the same probability, a multiplicity correction is applied in a natural way. In particular, the prior probability of finding at least one different treatment group will be exactly $1/(\zeta_0 + 1)$ no matter the number of treatment groups.

2.2 A Prior for the Group Distributions

As in De Iorio et al. [8], we set a dependent Dirichlet process prior for Q_j [20]:

$$Q_j = \sum_{k \in \mathbb{N}} w_k \mathrm{N}(\boldsymbol{\mu}_{jk}, \boldsymbol{\Sigma}_{jk}), \qquad \forall j \in \mathscr{J},$$

where

$$\begin{aligned}(\boldsymbol{\mu}_{jk}, \boldsymbol{\Sigma}_{jk}) &\overset{iid}{\sim} \mathrm{NIW}(\boldsymbol{u}_0, r_0, \nu_0, \boldsymbol{S}_0),\\ w_k &= v_k \prod\nolimits_{h=1}^{k-1} (1 - v_h),\\ v_k &\overset{iid}{\sim} \mathrm{B}(1, \alpha),\\ \alpha &\sim \mathrm{Ga}(a_0, b_0),\end{aligned} \tag{2}$$

and N(·), NIW(·), B(·), Ga(·) represent the Normal, Normal-Inverse-Wishart, Beta, and Gamma distributions, respectively (we will use the shape/rate parametrization for the Gamma distribution). Note that all the group distributions share the same weights. This assumption will simplify the posterior inference considerably.

3 Posterior Inference

To get the posterior $p(\boldsymbol{\gamma} \mid \boldsymbol{y})$ in (1) we need to calculate $L(\boldsymbol{y} \mid \boldsymbol{\gamma})$, which in turn requires marginalizing any additional variable in the model. In this case, the additional variables are the infinite dimensional objects (Q_j), which make the direct computation of $L(\boldsymbol{y} \mid \boldsymbol{\gamma})$ nearly impossible. In order to solve this problem, we propose to approximate $p(\boldsymbol{\gamma} \mid \boldsymbol{y})$ through a Gibbs sampler [12, 13]; more precisely, a Metropolis-within-Gibbs algorithm, as we update γ using the Metropolis-Hastings algorithm [15].

Before introducing our Gibbs sampler, let us rewrite our model in simpler terms. First, note that given $\boldsymbol{\gamma}$, the model behaves as if the effective group indicator was not x_i but $z_i := \gamma_{x_i} x_i$, under the convention $\gamma_0 = 0$. Second, note that we can always rewrite $G := \prod_{j \in \mathcal{J}} G_j$ as

$$G = \sum_{k \in \mathbb{N}} w_k \delta_{\xi_k}, \qquad \xi_k \overset{iid}{\sim} \bar{G} := \prod_{j \in \mathcal{J}} \mathrm{NIW}(\boldsymbol{u}_0, r_0, \nu_0, \boldsymbol{S}_0),$$

with w_k and v_k defined as in (2), which is the stick-breaking representation of a Dirichlet process [27] with concentration parameter α and base distribution $\bar{G}$, a $\mathrm{DP}(\alpha, \bar{G})$ process. Hence, we can rewrite our full model as

$$\begin{aligned} \boldsymbol{y}_i \mid z_i, ((\boldsymbol{\mu}_{ij}, \boldsymbol{\Sigma}_{ij}))_{j \in \mathcal{J}} &\overset{ind}{\sim} \mathrm{N}_D(\boldsymbol{\mu}_{iz_i}, \boldsymbol{\Sigma}_{iz_i}), \\ ((\boldsymbol{\mu}_{ij}, \boldsymbol{\Sigma}_{ij}))_{j \in \mathcal{J}} &\overset{iid}{\sim} G, \\ G &\sim \mathrm{DP}(\alpha, \bar{G}), \\ z_i &= \gamma_{x_i} x_i, \\ \boldsymbol{\gamma} &\sim \pi_0(\boldsymbol{\gamma}), \\ \alpha &\sim \mathrm{Ga}(a_0, b_0), \end{aligned} \tag{3}$$

and as G is a.s. discrete, the data will be clusterized. Indeed, let $((\boldsymbol{\mu}^\star_{ij}, \boldsymbol{\Sigma}_{ij})^\star)_{j \in \mathcal{J}}$ be the unique values in this sequence, then we can introduce N cluster membership indicators $s_1, \ldots, s_N \in [K]$ such that $s_i = k$ if and only if $\boldsymbol{\theta}_i = \boldsymbol{\theta}^\star_k$.

Now we are ready to explain our Gibbs sampler. Our algorithm is based on two key insights. The first one is that, given $\boldsymbol{\gamma}$, our model behaves as the multivariate counterpart of an ANOVA-DDP [8]. For this model, there are well-known procedures for updating (s_i), such as Neal's algorithm 3 [24]. The other insight is that, given

(z, s), our model behaves as a 2-way MANOVA model with z and s as factors:

$$\begin{aligned} y_i \mid z_i = j, s_i = k, \mu^\star_{jk}, \Sigma^\star_{jk} &\overset{ind}{\sim} \mathrm{N}_D(\mu^\star_{jk}, \Sigma^\star_{jk}), \\ (\mu^\star_{jk}, \Sigma^\star_{jk}) &\overset{iid}{\sim} NIW(u_0, r_0, \nu_0, S_0). \end{aligned}$$

Thus, conditionally on (γ, s), $p(y \mid z, s)$ factorizes as

$$p(y \mid z, s) = \prod_{j \in \mathscr{J}} \prod_{k \in [K]} p(\{y_i : i \in \mathscr{I}_{jk}\} \mid \mathscr{I}_{jk}),$$

where $\mathscr{I}_{jk} := \{i : z_i = j, s_i = k\}$. Moreover, $p(\{y_i : i \in \mathscr{I}_{jk}\} \mid \mathscr{I}_{jk})$ coincides with the likelihood of $\{y_i : i \in \mathscr{I}_{jk}\}$ under the model $y_i \mid \mu^\star_{jk}, \Sigma^\star_{jk} \sim \mathrm{N}_D(\mu^\star_{jk}, \Sigma^\star_{jk})$, $(\mu^\star_{jk}, \Sigma^\star_{jk}) \sim NIW(u_0, r_0, \nu_0, S_0)$. Hence, $p(\{y_i : i \in \mathscr{I}_{jk}\} \mid \mathscr{I}_{jk})$ can be treated analytically using the very well-known equation [2]:

$$p(\{y_i : i \in \mathscr{I}_{jk}\} \mid \mathscr{I}_{jk}) = \frac{1}{\pi^{Dn_{jk}/2}} \left(\frac{r_0}{r_{jk}}\right)^{D/2} \frac{|S_0|^{\nu_0/2}}{|S_{jk}|^{\nu_{jk}/2}} \frac{\prod_{d=1}^D \Gamma((\nu_{jk} + d - 1)/2)}{\prod_{d=1}^D \Gamma((\nu_0 + d - 1)/2)},$$

where $n_{jk} = \#\mathscr{I}_{jk}$, $\nu_{jk} = \nu_0 + n_{jk}$, $r_{jk} = r_0 + n_{jk}$, $u_{jk} = (r_0 u_0 + \sum_{i \in \mathscr{I}_{jk}} y_i)/r_{jk}$, and $S_{jk} = S_0 + \sum_{i \in \mathscr{I}_{jk}} y_i y_i' + r_0 u_0 u_0' - r_{jk} u_{jk} u_{jk}'$. Using this formula, we can compute (up to a proportionality constant) the value of $p(\gamma \mid y, x, s, \alpha) \propto p(y \mid x, s, \gamma, \alpha)\pi_0(\gamma)$.

Then, we propose a Gibbs sampler composed of three steps:

1. Draw α from its full conditional distribution using the method described in Escobar and West [9].
2. For each observation index $i \in [N]$, draw s_i from its full conditional distribution using Neal's algorithm 3 [24].
3. Draw $\gamma \sim p(\gamma \mid \alpha, s, x, y)$ through a Metropolis-Hastings algorithm using $K(\beta \mid \gamma) \propto I(\|\beta - \gamma\|_1 = 1)$ as a proposal distribution.

Once we get a sample from this Gibbs sampler, we can approximate $p(\gamma \mid y)$ using the empirical distribution of the sampled γ's. From $p(\gamma \mid y)$, we can recover the Bayes factors (if desired) from the equation

$$B_{\gamma,\beta} = \frac{L(\gamma \mid y)}{L(\beta \mid y)} = \frac{p(\gamma, y)}{p(\beta, y)} \frac{\pi_0(\beta)}{\pi_0(\gamma)} = \frac{p(\gamma \mid y)/p(y)}{p(\beta \mid y)/p(y)} \frac{\pi_0(\beta)}{\pi_0(\gamma)} = \frac{p(\gamma \mid y)}{p(\beta \mid y)} \frac{\pi_0(\beta)}{\pi_0(\gamma)}.$$

The full procedure is implemented in a Julia package, available at

https://github.com/igutierrezm/MANOVABNPTest.jl

The package can also be used from R using the `JuliaConnectoR` package. See the aforementioned repository for more details.

4 Monte Carlo Simulation Study

In order to evaluate the performance of our hypothesis testing procedure, we run a Monte Carlo experiment comparing our method with the test proposed by Mukhopadhyay and Wang [22]. We compare both methodologies on 72 different scenarios, resulting from the combination of three different sample sizes (indexed by $h = 1, 2, 3$), three different levels of similarity among the groups' distributions (indexed by $l = 1, 2, 3$), and eight different true hypotheses (indexed by $\boldsymbol{\beta} \in \{0, 1\}^3$). Specifically, for each $(h, l, \boldsymbol{\beta}) \in [3] \times [3] \times \{0, 1\}^3$, the data generating process (DGP) is

$$x_i = (i - 1) \bmod 4,$$

$$\boldsymbol{y}_i | x_i \sim \begin{cases} \mathrm{N}_2(\boldsymbol{0}_2, \boldsymbol{\Sigma}), & \text{if } x_i = 0, \\ \mathrm{N}_2(\beta_1 c_{l1} \boldsymbol{1}_2, \boldsymbol{\Sigma}), & \text{if } x_i = 1, \\ \mathrm{N}_2(\boldsymbol{0}_2, c_{l2}^{\beta_2} \boldsymbol{\Sigma}), & \text{if } x_i = 2, \\ 0.5\mathrm{N}_2(-\beta_3 c_{l3} \boldsymbol{1}_2, \boldsymbol{\Sigma}) + 0.5\mathrm{N}_2(\beta_3 c_{l3} \boldsymbol{1}_2, \boldsymbol{\Sigma}), & \text{if } x_i = 3, \end{cases} \qquad i \in (m_h),$$

where

$$\boldsymbol{\Sigma} = \begin{pmatrix} 1 & 0.3 \\ 0.3 & 1 \end{pmatrix}, \quad \begin{pmatrix} m_1 \\ m_2 \\ m_3 \end{pmatrix} = \begin{pmatrix} 200 \\ 600 \\ 1200 \end{pmatrix}, \quad \begin{pmatrix} c_{11} & c_{12} & c_{13} \\ c_{21} & c_{22} & c_{23} \\ c_{31} & c_{32} & c_{33} \end{pmatrix} = \begin{pmatrix} 0.625 & 1.498 & 1.162 \\ 0.819 & 1.806 & 1.359 \\ 1.051 & 1.994 & 1.652 \end{pmatrix}.$$

Note that each treatment group can differ in a different way from the control group. For example, group 1 has a difference in location, group 2 has a difference in scale, and group 3 has a distribution that does not translate into a difference in location. Note also the role of l: the bigger the l, the stronger is the difference across each group and the control group.

For each $(h, l, \boldsymbol{\beta})$, we evaluate the performance of our method as follows. First, we simulate 100 samples from the aforementioned DGP, and standardized the outcomes. Then, for each simulated sample, we approximate $p(\boldsymbol{\gamma} \mid \boldsymbol{y})$ using our approach and average the results over the 100 simulations. In order to implement our approach, we need to set the hyperparameters. Following Gutiérrez et al. [14], we set $(a_0, b_0, \zeta_0) = (1, 1, 1)$, and following Bouchard-Côté et al. [2], we set $(r_0, \nu_0, \boldsymbol{u}_0, \boldsymbol{S}_0) = (1, D + 2, \boldsymbol{0}_D, \boldsymbol{I}_D)$. By repeating these specifications for each $(h, l, \boldsymbol{\beta})$, we get $8 \times 72 = 576$ (averaged) posterior probabilities, one per each $(h, l, \boldsymbol{\beta}, \boldsymbol{\gamma}) \in [3] \times [3] \times \{0, 1\}^3 \times \{0, 1\}^3$.

Figure 2 displays these 576 numbers as a panel of heatmaps in greyscale, with $\boldsymbol{\beta}$ on the x-axis and $\boldsymbol{\gamma}$ on the y-axis of each plot, both arranged in the same order. In this heatmap, black represents 1 and white represents 0. Hence, for the perfect test, each element in the diagonal should be colored as black, and each element outside the diagonal should be colored as white. As expected, our procedure does not produce this perfect output. However, except for the case where we use the smallest sample

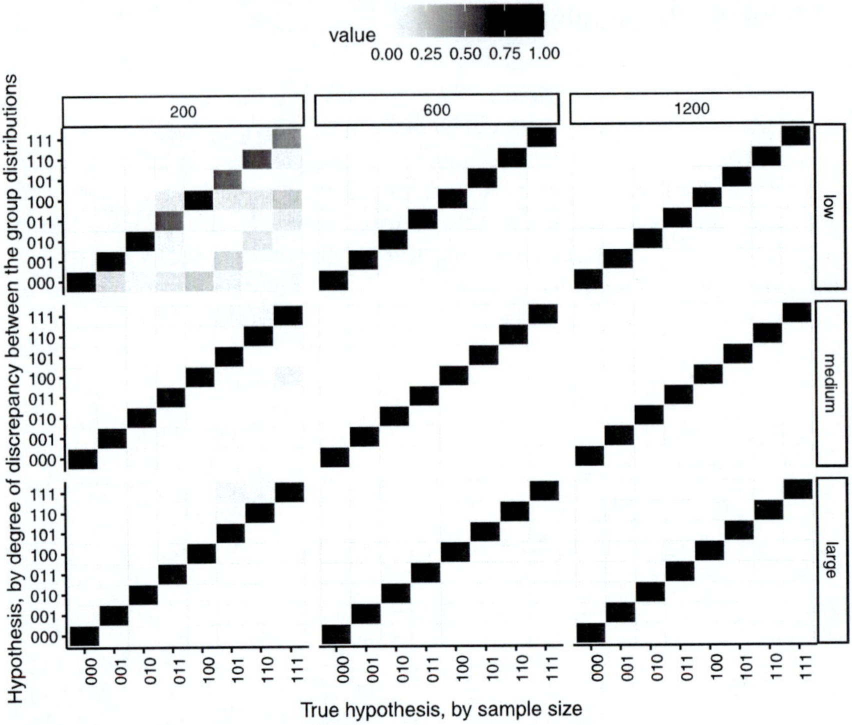

Fig. 2 Posterior probabilities (averaged over 100 runs) for the Monte Carlo experiment. For each sample, we run our Gibbs sampler for 4000 iterations and discard the first 2000. We standardize the responses before applying our method, with $(r_0, \nu_0, \boldsymbol{u}_0, \boldsymbol{S}_0, r_0, a_0, b_0, \zeta_0) = (1, D + 2, \boldsymbol{0}_D, \boldsymbol{I}_D, 1, 1, 1)$.

size and least obvious groups' differences, our proposal works well. In particular, the most probable hypothesis (on average) always coincided with the true hypothesis.

For the competitor, we used the same simulated samples. However, instead of approximating $p(\boldsymbol{\gamma} \mid \boldsymbol{y})$, we computed the hypothesis selected by the Mukhopadhyay and Wang [22] test (including Bonferroni correction), and then averaged the results for each $(h, l, \boldsymbol{\beta})$. Figure 3 displays these 576 relative frequencies as a second panel of heatmap, specified as the first one. The test worked almost as well as our proposal for large samples. However, for small samples sizes, this frequentist test is outerperformed by our BNP procedure.

Admittedly, the comparison is not completely fair because in Fig. 2 we are averaging posterior distributions of γ, whereas in Fig. 3 we are averaging concrete estimations of γ. However, for this particular simulation example, using a Bayesian estimator of γ in Fig. 2 (e.g., its MAP estimator) would not change the results ina significant way, specially on panels (2, 1) and (3, 1), where the difference between the performances is obvious.

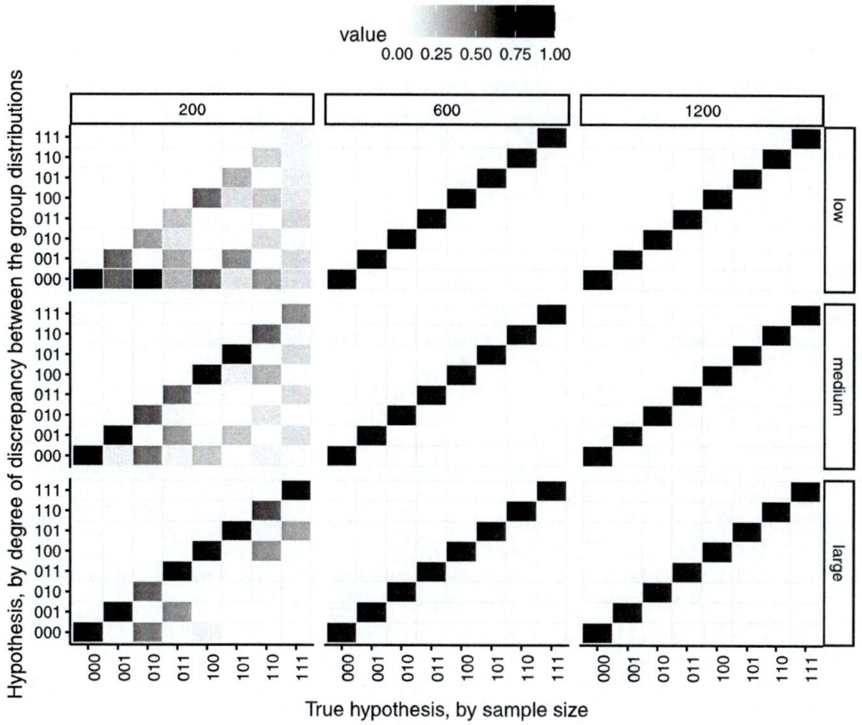

Fig. 3 Distribution of the hypothesis selected by the Mukhopadhyay and Wang [22] test (with Bonferroni correction) across 100 simulations

5 Discussion

In this article we introduced a Bayesian nonparametric testing procedure for comparing several treatment groups against some pre-specified control group. Our Monte Carlo experiments suggest that our model can successfully detect difference in location, scale and distribution. In addition, our procedure works similarly or better than state-of-the-art nonparametric k-sample tests. In comparison to other classical proposals, our hypothesis testing applies a multiplicity correction in a more principled way, and in comparison to other BNP approaches, our proposal is easier to apply in multivariate settings (in fact, many parameters can be marginalized), easier to tune (since we use Dirac spike-and-slab priors instead of absolutely continuous) and easier to communicate (since our construction is based on the well-known Dirichlet process).

As future work we highlight potential improvements to the Gibbs sampler described in Sect. 3. For example, we could use a split-merge procedure [17] instead of Neal's algorithm 3, which is relatively more efficient. Also, we could update $\boldsymbol{\gamma}$ using an informed proposal as the one described in Zanella [33]. All these improvements would be specially relevant for applications with a large number of observations and groups. In addition, we highlight the possibility of using even more flexible models for the involved group distributions. Indeed, under our construction, the weights are entirely shared by the groups, but it is also possible to generate weights that are only partially shared across the groups [29].

References

1. Blackwell, D., MacQueen, J.B.: Ferguson distributions via Polya urn schemes. Ann. Stat. **1**(2), 353–355 (1973)
2. Bouchard-Côté, A., Doucet, A., Roth, A.: Particle Gibbs split-merge sampling for Bayesian inference in mixture models. J. Mach. Learn. Res. **18**(28), 1–39 (2017)
3. Chatfield, C., Collins, A.J.: Introduction to Multivariate Analysis. Chapman and Hall (1980)
4. Chen, H., Chen, X., Su, Y.: A weighted edge-count two-sample test for multivariate and object data. J. Am. Stat. Assoc. **113**(523), 1146–1155 (2018)
5. Chen, H., Friedman, J.H.: A new graph-based two-sample test for multivariate and object data. J. Am. Stat. Assoc. **112**(517), 397–409 (2017)
6. Chen, Y., Hanson, T.E.: Bayesian nonparametric k-sample tests for censored and uncensored data. Comput. Stat. Data Anal. **71**, 335–346 (2014)
7. Cipolli, W., III., Hanson, T.E., McLain, A.C.: Bayesian nonparametric multiple testing. Comput. Stat. Data Anal. **101**, 64–79 (2016)
8. De Iorio, M., Müller, P., Rosner, G.L., MacEachern, S.N.: An ANOVA model for dependent random measures. J. Am. Stat. Assoc. **99**(465), 205–215 (2004)
9. Escobar, M.D., West, M.: Bayesian density estimation and inference using mixtures. J. Am. Stat. Assoc. **90**(430), 577–588 (1995)
10. Ferguson, T.S.: A Bayesian analysis of some nonparametric problems. Ann. Stat. **1**(2), 209–230 (1973)
11. Friedman, J.H., Rafsky, L.C.: Multivariate generalizations of the Wald-Wolfowitz and Smirnov two-sample tests. Ann. Stat. **7**(4), 697–717 (1979)
12. Gelfand, A.E., Smith, A.F.M.: Sampling-based approaches to calculating marginal densities. J. Am. Stat. Assoc. **85**(410), 398–409 (1990)
13. Geman, S., Geman, D.: Stochastic relaxation, Gibbs distribution, and the Bayesian restoration of images. IEEE Trans. Pattern Anal. Mach. Intell. **6**, 721–741 (1984)
14. Gutiérrez, L., Barrientos, A.F., Gonzalez, J., Taylor-Rodríguez, D.: A Bayesian nonparametric multiple testing procedure for comparing several treatments against a control. Bayesian Anal. **14**(2), 649–675 (2019)
15. Hastings, W.K.: Monte Carlo sampling methods using Markov chains and their applications. Biometrika **57**, 97–109 (1970)
16. Holmes, C.C., Caron, F., Griffin, J.E., Stephens, D.A.: Two-sample Bayesian nonparametric hypothesis testing. Bayesian Anal. **10**(2), 297–320 (2015)
17. Jain, S., Neal, R.M.: A split-merge Markov chain Monte Carlo procedure for the Dirichlet process mixture model. J. Comput. Graph. Stat. **13**(1), 158–182 (2004)
18. Jefferys, W.H., Berger, J.O.: Ockham's razor and Bayesian analysis. Am. Sci. **80**(1), 64–72 (1992)
19. Ma, L., Wong, W.H.: Coupling optional Pólya trees and the two sample problem. J. Am. Stat. Assoc. **106**(496), 1553–1565 (2011)
20. MacEachern, S.N.: Dependent nonparametric processes. ASA Proc. Sect. Bayesian Stat. Sci. Am. Statistical Association, Alexandria, VA (1999)
21. Mitchell, T.J., Beauchamp, J.J.: Bayesian variable selection in linear regression. J. Am. Stat. Assoc. **83**(404), 1023–1032 (1988)

22. Mukhopadhyay, S., Wang, K.: A nonparametric approach to high-dimensional k-sample comparison problems. Biometrika **107**(3), 555–572 (2020)
23. Müller, P., Quintana, F., Rosner, G.: A method for combining inference across related nonparametric Bayesian models. J. R. Stat. Soc. Ser. B Stat. Methodol. **66**(3), 735–749 (2004)
24. Neal, R.M.: Markov chain sampling methods for Dirichlet process mixture models. J. Comput. Graph. Stat. **9**(2), 249–265 (2000)
25. Rosenbaum, P.R.: An exact distribution-free test comparing two multivariate distributions based on adjacency. J. R. Stat. Soc. Ser. B Stat. Methodol. **67**(4), 515–530 (2005)
26. Scott, J.G., Berger, J.O.: An exploration of aspects of Bayesian multiple testing. J. Stat. Plan. Inference **136**(7), 2144–2162 (2006)
27. Sethuraman, J.: A constructive definition of Dirichlet priors. Stat. Sin. **4**(2), 639–650 (1994)
28. Soriano, J., Ma, L.: Probabilistic multi-resolution scanning for two-sample differences. J. R. Stat. Soc. Ser. B Stat. Methodol. **79**(2), 547–572 (2017)
29. Soriano, J., Ma, L.: Mixture modeling on related Samples by ψ-stick breaking and kernel perturbation. Bayesian Anal. **14**(1), 161–180 (2019)
30. Teh, Y., Jordan, M., Beal, M., Blei, D.: Hierarchical Dirichlet processes. J. Am. Stat. Assoc. **101**(476), 1566–1581 (2006)
31. Weiss, L.: Two-sample tests for multivariate distributions. Ann. Math. Stat. **31**(1), 159–164 (1960)
32. Womack, A.J., Fuentes, C., Taylor-Rodriguez, D.: Model Space Priors for Objective Sparse Bayesian Regression. arXiv e-prints p. arxiv:1511.04745 (2015)
33. Zanella, G.: Informed proposals for local MCMC in discrete spaces. J. Am. Stat. Assoc. **115**(530), 852–865 (2020)

Specification of the Base Measure of Nonparametric Priors via Random Means

Francesco Gaffi, Antonio Lijoi, and Igor Prünster

Abstract Functionals of random probability measures are probabilistic objects whose properties are studied in different fields. They also play an important role in Bayesian Nonparametrics: understanding the behavior of a finite dimensional feature of a flexible and infinite-dimensional prior is crucial for prior elicitation. In particular distributions of means of nonparametric priors have been the object of thorough investigation in the literature. We target the inverse path: the determination of the parameter measure of a random probability measure giving rise to a fixed mean distribution. This direction yields a better understanding of the sets of mean distributions of notable nonparametric priors, giving moreover a way to directly enforce prior information, without losing inferential power. Here we summarize and report results obtained in [6] for the Dirichlet process, the normalized stable random measure and the Pitman–Yor process, with an application to mixture models.

Keywords Random probability measures · Random means · Nonparametric prior elicitation · Dirichlet process · Pitman-Yor process · Normalized stable process

1 Introduction

Distributional properties of linear functionals of random probability measures (RPM) of the form

$$M_f(\tilde{P}) := \int_{\mathbb{X}} f(x)\tilde{P}(\mathrm{d}x) \quad \text{for } f : \mathbb{X} \to \mathbb{R} \text{ measurable} \tag{1}$$

F. Gaffi (✉) · A. Lijoi · I. Prünster
Department of Decision Sciences and BIDSA, Bocconi University, Milan, Italy
e-mail: francesco.gaffi@phd.unibocconi.it

A. Lijoi
e-mail: antonio.lijoi@unibocconi.it

I. Prünster
e-mail: igor@unibocconi.it

R. Argiento et al. (eds.), *New Frontiers in Bayesian Statistics*, Springer Proceedings in Mathematics & Statistics 405, https://doi.org/10.1007/978-3-031-16427-9_9

with $\mathbb{X}$ a Polish space and $\tilde{P}$ a RPM on $\mathbb{X}$, were studied in the seminal contributions by Cifarelli and Regazzini [1, 2] to tackle inferential problems in Bayesian Nonparametrics. This area of research generated many interesting results from different perspectives. See [29] for early results, [3] for a contextualization in the Bayesian inferential framework and beyond, and [15] for an exhasutive account of results and for a detailed account of the connections with other research fields. The distribution of $M(\tilde{P}) := \int_{\mathbb{R}} x\, \tilde{P}(\mathrm{d}x)$, with $\tilde{P}$ a RPM on $\mathbb{R}$, has been characterized for:

1 $\tilde{P} = \tilde{\mathscr{D}}_\alpha$ a Dirichlet process (DP) with parameter measure $\alpha = \theta P_0$
2 $\tilde{P} = \tilde{P}_{\sigma,\theta}$ a Pitman–Yor process (PYP) with parameters $\sigma \in (0, 1)$ and $\theta > -\sigma$
3 $\tilde{P}$ a *normalized random measure with independent increments* (NRMI)

respectively in [2, 16], [8] and [9, 23]. The results are obtained for $M(\tilde{P})$, as they can be extended to a linear functional $M_f(\tilde{P})$, for any measurable function $f : \mathbb{X} \to \mathbb{R}$ such that $\int |f|\, \mathrm{d}\tilde{P} < \infty$, almost surely. This follows from

$$\int f\, \mathrm{d}\tilde{P} \overset{d}{=} \int x\, \tilde{P}_f(\mathrm{d}x) \tag{2}$$

where $\tilde{P}_f = \tilde{P} \circ f^{-1}$. As far as Dirichlet random means are concerned, a preliminary result in [5] gives a necessary and sufficient condition for (1) to exist almost surely finite when $\tilde{P} = \tilde{\mathscr{D}}_\alpha$, namely

$$\int_{\mathbb{X}} \log(1 + |f(x)|)\, \alpha(\mathrm{d}x) < \infty. \tag{3}$$

In [2] a first characterization and the absolute continuity of the law of a random Dirichlet mean are proved, while in [16] explicit expressions for the density function and the characteristic function are given. Moreover, [16] underlines, uses and extends the connection of Dirichlet random densities with Lauricella hypergeometric functions. Furthermore, general assumptions that allow to attain an explicit expression of the mean density and the mean cdf, as well as results on symmetry of the mean distribution and on vectors of Dirichlet random means, can be found in [16] and in [22].

Turning attention to means of a PYP instead, in [23] we find the almost sure finiteness condition for $M_f(\tilde{P}_{\sigma,\theta})$, that is

$$\int |f|^\sigma\, \mathrm{d}P_0 < \infty \tag{4}$$

where $P_0 = \mathbb{E}\left[\tilde{P}_{\sigma,\theta}\right]$. In [8] an explicit expression for the density function of $M(\tilde{P}_{\sigma,\theta})$, together with the proof of the absolute continuity of its law, is presented. Moreover, in [15] we can find an account of the connection between PYP means and excursions of skew Bessel bridges, as introduced in [19].

In [23] results for general NRMI means are proved. A NRMI on a Polish space $\mathbb{X}$ can be defined as

$$\tilde{P}(\cdot) = \frac{\tilde{\mu}(\cdot)}{\tilde{\mu}(\mathbb{X})} \tag{5}$$

where $\tilde{\mu}$ is a proper *completely random measure* (CRM) on $\mathbb{X}$ (namely without fixed locations) such that $0 < \tilde{\mu}(\mathbb{X}) < \infty$ almost surely. As described in [14], $\tilde{\mu}$ is characterized by the *Laplace functional*

$$\mathbb{E}\left[e^{-\int_{\mathbb{X}} f(x)\,\tilde{\mu}(dx)} \right] = \exp\left\{ -\int_{\mathbb{R}^+\times\mathbb{X}} \left[1 - e^{-sf(x)}\right] \nu(ds, dx) \right\} \tag{6}$$

where $f : \mathbb{X} \to \mathbb{R}$ is a measurable function such that $\int |f|\, d\tilde{\mu} < \infty$ (almost surely) and ν is a measure on $\mathbb{R}^+ \times \mathbb{X}$ such that

$$\int_B \int_{\mathbb{R}^+} \min\{s, 1\}\, \nu(ds, dx) < \infty \tag{7}$$

for any B in $\mathscr{B}(\mathbb{X})$. The measure ν is named the Lévy intensity of $\tilde{\mu}$. In this case the random mean $M_f(\tilde{P})$ exists almost surely finite if and only if

$$\int_{\mathbb{X}\times\mathbb{R}^+} (1 - e^{-\lambda v |f(x)|})\nu(dv,\, dx) < \infty \quad \text{for every} \quad \lambda > 0. \tag{8}$$

A general expression for the cumulative distribution of a NRMI mean is given in [23].

A crucial analytical tool in obtaining these results is the *generalized Cauchy–Stieltjes transform*

$$\text{for } g : \mathbb{R}^+ \to \mathbb{R} \qquad \mathscr{S}_\lambda[z; g] := \int_{\mathbb{R}^+} \frac{g(x)}{(z+x)^\lambda}\, dx \qquad \lambda > 0,\ z \in \mathbb{C} \setminus \mathbb{R}^- \tag{9}$$

whose properties and inversion formulas can be found in [10, 25, 26]. It is involved in *Cifarelli–Regazzini* (CR) *identities*, which equate integral transforms of base measures and of the corresponding mean density distributions. The original CR identity, proved in [2, 27], concerns the Dirichlet case and states the following

$$\mathscr{S}_\theta[z;\, q_\alpha] = \exp\left\{ -\int_{\mathbb{R}} \log(z+x)\, \alpha(dx) \right\} \qquad z \in \mathbb{C} \setminus \mathbb{R}^- \tag{10}$$

where q_α is the mean density function of $M(\tilde{\mathscr{D}}_\alpha)$. A generalized version, proved in [11, 27, 28], holds for the PYP case: for any $\sigma \in (0, 1)$ and $\theta > 0$

$$\mathscr{S}_\theta\left[z;\, q_{\sigma,\theta}\right] = \left\{ \int_{\mathbb{R}} (z+x)^\sigma \, P_0(\mathrm{d}x) \right\}^{-\theta/\sigma} \qquad z \in \mathbb{C} \setminus \mathbb{R}^- \tag{11}$$

where $q_{\sigma,\theta}$ is the mean density function of $M(\tilde{P}_{\sigma,\theta})$ and $P_0 = \mathbb{E}\left[\tilde{P}_{\sigma,\theta}\right]$.

To sum up, the standard approach has been relying on CR identities in order to characterize random means distributions by, e.g., determining their density functions, given the base measure P_0. The purpose of our work is to address the inverse problem, which amounts to determining the base measure of a RPM $\tilde{P}$ leading to a specific distribution for $M(\tilde{P})$. Pursuing this direction has a strong motivation in Bayesian inference: if one has enough *a priori* information to elicit the distribution of an interpretable finite-dimensional feature of $\tilde{P}$, such as its mean, then it is crucial to identify which parameters for $\tilde{P}$ enforce correctly the prior knowledge we have on such a feature. Moreover, our results allow to induce the elicited distribution on the mean without otherwise modifying the random measure, that is keeping its distribution in a selected class (e.g. DP or PYP). This entails that posterior inference, whenever it is available for such class of nonparametric priors, can be leveraged on with no further adjustments.

The solution to a special case of such a problem can be deduced from results in the combinatorial literature, and this constitutes a further instance of connection with seemingly unrelated research areas. In particular from [24] we can extrapolate an expression for the cumulative distribution function (cdf) of the base measure of a DP with concentration parameter $\theta = 1$. In [6] results are provided yielding the base measure corresponding to a fixed mean distribution for a DP with $\theta < 1$, for a $\sigma-$stable NRMI and for a PYP. Here we present the main statements and the statistical implications of their application to mixture models.

2 Dirichlet Case

In the DP case, the CR identity (10) ensures that the total mass $\theta > 0$ and the mean distribution Q_α uniquely identify the base measure. Hence, once θ is fixed, it is a well-posed question to ask which $P_0 = \frac{\alpha}{\theta}$ induces a desired distribution q_α on a Dirichlet random mean.

This problem has been addressed, from a completely different perspective, in [24] by exploring the relationship between *continual Young diagrams* and the transition measure they induce on a compact interval via a Markov process known as *hook walk*. An early account on such combinatorial objects is given in [11]. Surprisingly, if one considers the proper subset of convex diagrams, such relationship is the same linking the base measure and the mean distribution of a DP when $\theta = 1$. Hence, leveraging on results in [24], we obtain

$$P_0([0,x]) = \alpha([0,x]) = \frac{1}{\pi}\operatorname{arccot}\left(\frac{1}{\pi\, q(x)}\,\mathrm{PV}\int_0^1 \frac{q(t)}{t-x}\,\mathrm{d}t\right) \tag{12}$$

with q being the DP mean density and PV $\int$ denoting the *Cauchy principal value*. For details about such singular integrals see e.g. [4]. Sufficient conditions for this to hold are that q is piecewise C^1, bounded away from 0 and with bounded derivative.

Let us now consider $\theta \in (0, 1)$. For a density function f such that

$$\int_0^1 \frac{f(x)}{|x-t|^\theta}\,\mathrm{d}x < \infty \qquad \forall t \in [0,1]$$

we define

$$\mathscr{I}_\theta[\,f;\,t\,] := \frac{\displaystyle\int_t^1 \frac{f(x)}{|x-t|^\theta}\,\mathrm{d}x}{\displaystyle\int_0^t \frac{f(x)}{|x-t|^\theta}\,\mathrm{d}x} \qquad t \in (0,1] \tag{13}$$

Then it is possible to state the following.

Theorem 1 *Let $\theta \in (0, 1)$ and q_α be the density of $M(\tilde{\mathscr{D}}_\alpha)$ with $\mathrm{supp}(q_\alpha) = [0, 1]$. If*

$$\int_0^1 \frac{q_\alpha(x)}{|x-t|^\theta}\,\mathrm{d}x < \infty \qquad \forall t \in [0,1] \tag{14}$$

then the cdf of the base measure P_0 is given by

$$F_0(t) = \frac{1}{\theta}\left\{\frac{1}{\pi}\arctan\left(\frac{\sin(\theta\pi)}{\cos(\theta\pi) + \mathscr{I}_\theta[\,q_\alpha;\,t\,]}\right) + \mathbb{1}_{(t_*,\infty)}(t)\right\}\mathbb{1}_{(0,1)}(t) + \mathbb{1}_{[1,\infty)}(t) \tag{15}$$

where

$$t_* = \inf\left\{t \in [0,1] \,\middle|\, \mathscr{I}_\theta[\,q_\alpha;\,t\,] \le -\cos(\theta\pi)\right\} \tag{16}$$

Note that the integrability condition (14), even if required *for any* t, is not restrictive, since $\theta \in (0, 1)$. Moreover, unlike in (12) for $\theta = 1$, we do not assume smoothness of the mean density. We also underline that the compact support hypothesis may not be considered that restrictive in view of nonparametric prior elicitation and could be anyhow relaxed as done e.g. in [7].

As an example, the cdf of the base measure inducing a uniform distribution on the random mean of a Dirichlet process is given in the following. For a specific concentration parameter, (15) boils down to a particularly simple expression.

Example 1 Let $q_\alpha(\cdot) = \mathbb{1}_{[0,1]}(\cdot)$ and $\theta = \frac{1}{2}$. Then

$$F_0(t) = \frac{2}{\pi} \arctan \sqrt{\frac{t}{1-t}} \qquad t \in (0,1)$$

3 Normalized Stable and Pitman–Yor Cases

Let α be a measure on $\mathbb{X}$ and $\sigma \in (0,1)$. A CRM $\tilde{\mu}_\sigma$ with Lévy intensity given by

$$\nu(\mathrm{d}v, \mathrm{d}x) = \frac{\sigma}{\Gamma(1-\sigma)} v^{-1-\sigma}\, \mathrm{d}v\, \alpha(\mathrm{d}x) \tag{17}$$

is a *σ-stable CRM* with parameter measure α on $\mathbb{X}$. If $\alpha(\mathbb{X}) = \theta < \infty$, one can consider

$$\tilde{P}_\sigma := \frac{\tilde{\mu}_\sigma}{\tilde{\mu}_\sigma(\mathbb{X})} \tag{18}$$

obtaining the *σ-stable NRMI*, as in [13]. In [6] it is shown that, unlike for DP, once fixed a probability measure P_0, any $\alpha = \theta P_0$ leads to the same law for $M(\tilde{P}_\sigma)$, regardless of θ. Hence we shall set $\theta = 1$. Consider $\alpha = P_0$ supported on [0, 1], then we can state the following.

Theorem 2 *Let the density q_σ of $M(\tilde{P}_\sigma)$ be piecewise Hölder continuous and such that*

$$\int_0^1 |\log|x-t||\; q_\sigma(x)\, \mathrm{d}x < \infty \tag{19}$$

Lebesgue-almost everywhere. Then the base measure P_0 has cdf given by

$$F_0(y) = \frac{1}{\pi} \int_0^y (y-t)^{-\sigma}\, \mathrm{e}^{\sigma \int_0^1 \log|x-t|\, q_\sigma(x)\, \mathrm{d}x} \left\{ \pi\, q_\sigma(t) \cos(\sigma\pi\, Q_\sigma(t)) + \sin(\sigma\pi\, Q_\sigma(t))\, \mathrm{PV}\!\int_0^1 \frac{q_\sigma(x)}{t-x}\, \mathrm{d}x \right\} \mathrm{d}t \tag{20}$$

for any $y \in (0,1)$, where Q_σ is the cdf of q_σ.

The key idea of the proof relies on the following limiting version of the generalized CR identity (11), proved in [27]

$$\exp\left\{-\int \log(z+x)\, q_\sigma(x)\,\mathrm{d}\,x\right\} = \left\{\int (z+x)^\sigma\, P_0(\mathrm{d}\,x)\right\}^{1/\sigma} \tag{21}$$

Arguments on existence and regularity of singular integrals are crucial in determining sufficient conditions on q_σ. Moreover, the piecewise Hölder hypothesis can be relaxed, by exploiting results from singular integral approximation literature reported in [17] and [21].

In [8] a representation of PYP means in terms of DP and σ–stable NRMI means is established. It reads

$$\int x\, \tilde{P}_{\sigma,\theta}(\mathrm{d}\,x) \overset{d}{=} \int x\, \tilde{\mathscr{D}}_\alpha(\mathrm{d}\,x) \tag{22}$$

where $\alpha(B) = \theta \int_B q_\sigma(x)\,\mathrm{d}x$ and $\mathbb{E}\left[\tilde{P}_{\sigma,\theta}\right] = P_0$. In words, a Pitman-Yor(σ, θ) random mean has the same distribution as a Dirichlet random mean whose base measure is given by θ times the mean distribution of a σ-stable NRMI. By combining (22), (12) and Theorem 2 one can obtain the following result for the PYP.

Theorem 3 *Let the density $q_{\sigma,1}$ of the mean $M(\tilde{P}_{\sigma,1})$ of a PYP with parameters $(\sigma, 1)$ be piecewise C^1 with piecewise Hölder continuous derivative. Then the base measure P_0 of $\tilde{P}_{\sigma,1}$ has cdf given by*

$$F_0(y) = \frac{1}{\pi}\int_0^y (y-t)^{-\sigma}\, e^{\sigma \int_0^1 \log|x-t|\, q_\sigma(x)\,\mathrm{d}\,x}$$
$$\left\{\pi\, q_\sigma(t)\, \cos(\sigma\pi\, Q_\sigma(t)) + \sin(\sigma\pi\, Q_\sigma(t))\,\mathrm{PV}\int_0^1 \frac{q_\sigma(x)}{t-x}\,\mathrm{d}\,x\right\}\mathrm{d}\,t \tag{23}$$

with q_σ having cdf given by

$$Q_\sigma(t) = \frac{1}{\pi}\,\mathrm{arccot}\left(\frac{1}{\pi\, q_{\sigma,1}(t)}\,\mathrm{PV}\int_0^1 \frac{q_{\sigma,1}(x)}{x-t}\,\mathrm{d}x\right) \tag{24}$$

The expression of sufficient conditions directly on $q_{\sigma,1}$ requires the assessment, achieved in [6], of a real CR identity, linking the mean density function of a PYP with concentration parameter $\theta = 1$ and the mean distribution function of the underling stable NRMI, where we refer to the derivation of the PYP as normalization of a power tilting of a stable CRM proposed in [20]. This identity reads

$$\frac{\cos(\pi\, Q_\sigma(t))}{1-t}\exp\left\{\mathrm{PV}\int_0^1 \frac{Q_\sigma(s)}{s-t}\,\mathrm{d}s\right\} = \mathrm{PV}\int_0^1 \frac{q_{\sigma,1}(s)}{s-t}\,\mathrm{d}s \tag{25}$$

for Q_σ being the cdf of a stable mean and $q_{\sigma,1}$ being the density of a PYP mean with $\theta = 1$.

Explicit expressions of the base measure corresponding to specific choices of the density of the mean, are identified in [6] through Theorems 1, 2 and 3. Moreover, some special cases not covered by the general results are treated and solved.

4 Application to Mixture Models

Letting $\mathbb{Y}$ be a Polish space, a ν-absolutely continuous random mixture density is

$$\tilde{p}(y) = \int_{\mathbb{X}} k(y, x)\, \tilde{P}(\mathrm{d}x) \tag{26}$$

where $\{k(\cdot\,, x) : x \in \mathbb{X}\}$ is a collection of density functions on $\mathbb{Y}$ indexed by a parameter in $\mathbb{X}$. One can consider a linear functional of the mixture (26) and be interested in fixing its distribution, as in [12]. For instance, in a model where data are $\tilde{p}$-distributed, conditionally on $\tilde{p}$, it is natural that one wants to enforce a prior belief on the distribution of the population mean

$$\int y\tilde{p}(y)\, \nu(\mathrm{d}y).$$

In order to use the described results on linear functionals of random probability measures in this modeling framework, one can notice that, for a function $f : \mathbb{Y} \longrightarrow \mathbb{R}$

$$\int_{\mathbb{Y}} f(y)\, \tilde{p}(y)\nu(\mathrm{d}y) = \int_{\mathbb{X}} g(x)\, \tilde{P}(\mathrm{d}x) \tag{27}$$

where $g(x) = \int_{\mathbb{Y}} f(y)k(y, x)\nu(\mathrm{d}y)$. This strategy has been applied in [18] for deriving the distribution of means of DP and NRMI mixtures and implies that means with respect to a random mixture density can be treated as means with respect to the mixing random probability measure. Hence, it follows that Theorems 1, 2 and 3 can be applied in this case to determine which parameter measure to use for mixing random measure to induce a desired distribution on the functional in (27). This yields an important tool for prior specification in mixture models.

Acknowledgements The authors are grateful to the Editor and two anonymous Referees for their insightful comments and suggestions.

References

1. Cifarelli, D.M., Regazzini, E.: A general approach to Bayesian analysis of nonparametric problems. Decis. Econ. Finance **2**(1), 39–52 (1979)
2. Cifarelli, D.M., Regazzini, E.: Distribution functions of means of a Dirichlet process. Ann. Stat. **18**(1), 429–442 (1990)
3. Diaconis, P., Kemperman, J.: Some new tools for Dirichlet priors. In: J. Bernardo, J. Berger, A. Dawid, A. Smith (eds.) Bayesian Statistics, vol. 5, pp. 97–106. Oxford University Press (1996)
4. Estrada, R., Kanwal, R.P.: Singular integral equations. Birkhäuser Boston (2012)
5. Feigin, P.D., Tweedie, R.L.: Linear functionals and Markov chains associated with Dirichlet processes. Math. Proc. Camb. Philos. Soc. **105**(3), 579–585 (1989)
6. Gaffi, F., Lijoi, A., Prünster, I.: Random probability measures with fixed mean distribution. Working Paper (2022+)
7. Guglielmi, A.: A simple procedure calculating the generalized Stieltjes transform of the mean of a Dirichlet process. Stat. Probab. Lett. **38**(4), 299–303 (1998)
8. James, L.F., Lijoi, A., Prünster, I.: Distributions of linear functionals of two parameter Poisson-Dirichlet random measures. Ann. Appl. Probab. **18**(2), 521–551 (2008)
9. James, L.F., Lijoi, A., Prünster, I.: On the posterior distribution of classes of random means. Bernoulli **16**(1), 155–180 (2010)
10. Karp, D., Prilepkina, E.: Generalized Stieltjes functions and their exact order. J. Class. Anal. **1**(1), 53–74 (2012)
11. Kerov, S.V., Tsilevich, N.: The Markov-Krein correspondence in several dimensions. J. Math. Sci. **121**(3), 2345–2359 (2004)
12. Kessler, D., Hoff, P., Dunson, D.: Marginally specified priors for nonparametric Bayesian estimation. J. R. Stat. Soc. Series B Stat. Methodol. **77**, 35–58 (2015)
13. Kingman, J.F.C.: Random discrete distributions. J. R. Stat. Soc. Series B Stat. Methodol. **37**(1), 1–22 (1975)
14. Kingman, J.F.C.: Poisson Processes. Oxford Studies in Probability. Oxford University Press, United Kingdom (1993)
15. Lijoi, A., Prünster, I.: Distributional properties of means of random probability measures. Stat. Surv. **3**, 47–95 (2009)
16. Lijoi, A., Regazzini, E.: Means of a Dirichlet process and multiple hypergeometric functions. Ann. Probab. **32**(2), 1469–1495 (2004)
17. Martin, P., Rizzo, F.: Hypersingular integrals: how smooth must the density be? Int. J. Numer. Methods Eng. **39**, 687–704 (1996)
18. Nieto-Barajas, L., Prünster, I., Walker, S.: Normalized random measures driven by increasing additive processes. Ann. Stat. **32**, 2343–2360 (2004)
19. Pitman, J.: On the relative lengths of excursions derived from a stable subordinator. In: Azéma, J., Yor, M., Émery, M. (eds.) Séminaire de probabilités XXXI, pp. 287–305. Springer, Berlin (1997)
20. Pitman, J., Yor, M.: The two-parameter Poisson-Dirichlet distribution derived from a stable subordinator. Ann. Probab. **25**(2), 855–900 (1997)
21. Rabinowitz, P.: Uniform convergence results for Cauchy principal value integrals. Math. Comput. **56**, 731–740 (1991)
22. Regazzini, E., Guglielmi, A., Di Nunno, G.: Theory and numerical analysis for exact distributions of functionals of a Dirichlet process. Ann. Stat. **30**(5), 1376–1411 (2002)
23. Regazzini, E., Lijoi, A., Prünster, I.: Distributional results for means of normalized random measures with independent increments. Ann. Stat. **31**(2), 560–585 (2003)
24. Romik, D.: Explicit formulas for hook walks on continual Young diagrams. Adv. Appl. Math. **32**(4), 625–654 (2004)
25. Schwarz, J.H.: The generalized Stieltjes transform and its inverse. J. Math. Phys. **46** (2005)
26. Sumner, D.B.: An inversion formula for the generalized Stieltjes transform. Bull. Am. Math. Soc. **55**, 174–183 (1949)

27. Tsilevich, N.: Distribution of the mean value for certain random measures. J. Math. Sci. **96**(5), 3616–3623 (1999)
28. Vershik, A.M., Yor, M., Tsilevich, N.: On the Markov-Krein identity and quasi-invariance of the gamma process. J. Math. Sci. **121**(3), 2303–2310 (2004)
29. Yamato, H.: Characteristic functions of means of distributions chosen from a Dirichlet process. Ann. Probab. **12**(1), 262–267 (1984)

Bayesian Nonparametric Predictive Modeling for Personalized Treatment Selection

Matteo Pedone, Raffaele Argiento, and Francesco C. Stingo

Abstract We develop a Bayesian nonparametric predictive model to establish personalized therapeutic strategies for oncology patients. We leverage characteristics of both the patient and disease to support decision making in the selection of the optimal treatment. The core component of the model is a product partition model with covariates (PPMx) that induces clusters of observations that are more homogeneous with respect to predictive biomarkers. We conduct a simulation study to evaluate different modeling choices regarding PPMx in the framework of personalized treatment selection.

Keywords Product partition models · Nonparametric Bayes · Model-based clustering · Personalized medicine

1 Introduction

Our approach is motivated by an open problem in cancer genomics and personalized medicine. Personalized medicine's mission is to tailor treatment to individual patient characteristics leveraging various sources of heterogeneity. The distinctive mark of statistical inference under the personalized medicine paradigm is to disregard heterogeneity as nuisance to inference, but rather to take advantage of it to improve therapeutic strategies [2]. Cancer is a complex process and, to understand underlying biological phenomena, heterogeneity in both patients and disease must be accounted.

M. Pedone (✉) · F. C. Stingo
Universitá degli Studi di Firenze, viale Morgagni 65, 50134 Firenze, Italy
e-mail: matteo.pedone@unifi.it

F. C. Stingo
e-mail: francescoclaudio.stingo@unifi.it

R. Argiento
Universitá degli Studi di Bergamo, via Salvecchio 19, 24129 Bergamo, Italy
e-mail: raffaele.argiento@unibg.it

R. Argiento et al. (eds.), *New Frontiers in Bayesian Statistics*, Springer Proceedings in Mathematics & Statistics 405, https://doi.org/10.1007/978-3-031-16427-9_10

We develop a method for personalized treatment selection that leverages prognostic and predictive biomarkers.

Prognostic biomarkers impact the likelihood of achieving a therapeutic response regardless of the selected treatment. By contrast, predictive biomarkers determine which patients are likely or unlikely to benefit from a particular class of treatment regimes. Since cancer is an inherently heterogeneous disease, each tumor is unique and hence, for predictive covariates, patients should not be regarded as exchangeable [3]. Given genomic signatures and a set of prognostic markers, building on [4] we leverage prognostic determinants to measure how likely a patient is to reach a given clinical response. Predictive biomarkers are exploited to drive patients clustering within each treatment. This is done to typify the extent of benefit offered by a specific therapeutic strategy on groups of patients characterized by close profiles in predictive determinants. We are assuming to know which biomarkers are prognostic and which are predictive. Although this assumption seems restrictive, it remains crucial. Biomarkers, in order to lead to optimal treatment selection, need to be validated on completely independent data set not used during development. That is, rather than develop prognostic/predictive biomarkers, our goal is personalized treatment selection employing validated biomarkers.

The Bayesian framework naturally handles model-based clustering assuming as random parameter of the model the partition of the sample subjects. In particular, we adopt the product partition model with covariates (PPMx) [5] to induce clusters of observations that are more homogeneous with respect to predictive covariates, building partitions that are only partially exchangeable. The class of PPMx models is a powerful Bayesian nonparametric tool to incorporate covariates' information into the prior for the random partition. Indeed, under this class of models, patients with similar covariates are a priori more likely to be clustered together. This feature enables us to quantify the effectiveness of each competing therapeutic strategy for patients with similar genetic profiles.

Finally, the posterior predictive distribution of this model arises as a natural way to assess the extent to which a new untreated patient is likely to attain a level of clinical response for competing treatments. We elicit response utility weights and evaluate utility expectation for each therapy [3]. The treatment with the largest mean predictive utility is considered the optimal treatment.

The goal of this paper is to provide guidance regarding the specification of the prior distribution for the random partition in the framework of optimal treatment selection. In fact, as the number of predictive biomarkers grows, the influence of PPMx models on clustering tends to overwhelm the information from the response, negatively affecting inference and out-of-sample prediction. In order to calibrate the influence that covariates have on partition probabilities we follow [8]'s strategy to temper covariate impact on clustering. The evaluation of different calibrations is empirically done through simulations based on gene expression data from a leukemia study [1].

The remainder of the article is organized as follows. In Sect. 2 we state the proposed model, focusing on the aspects addressed in the simulation study. We also give some details on the computational strategy. In Sect. 3 we describe the predictive utility approach adopted for treatment selection. We report and discuss the results of the simulation study in Sect. 4 and Sect. 5 concludes the paper.

2 The Model

Let $a = 1, \ldots, T$ index candidate therapies to whom $n = \sum_{a=1}^{T} n^a$ patients are assigned to, where n^a denotes the number of patients treated with therapy a. A common choice to characterize varying levels of treatment response is to evaluate it in terms of the extent of residual disease after a given clinically relevant post-therapy follow-up duration. Let y_i^a be the random variable of the $i-$th patient's response to treatment a among K possible levels of increasing treatment benefit, where $y_i^a = k$ for $i = 1, \ldots, n^a$ and $k = 1, \ldots, K$. In addition, let $\boldsymbol{\pi}_i^a = (\pi_{i1}^a, \ldots, \pi_{iK}^a)$ denote the vector such that π_{ik}^a is the probability of observing outcome k for the $i-$th patient under treatment a. The treatment response is an ordinal-valued random variable and y_i^a follows a multinomial distribution $y_i^a \mid \boldsymbol{\pi}_i^a \stackrel{\text{ind}}{\sim} \text{Multinomial}(1, \boldsymbol{\pi}_i^a)$. For each treatment, we consider a training dataset of n^a patients, $(y_i^a, z_i^a, \boldsymbol{x}_i^a)$ where $i = 1, \ldots, n^a$ and z_i^a and $\boldsymbol{x}_i^a$ are a $P-$dimensional and $Q-$dimensional vector of prognostic and predictive features, respectively.

As mentioned in Sect. 1, to relax exchangeability among observations, we adopt a model for random partition depending on predictive markers. We denote with $\rho^a = \{S_1^a, \ldots, S_{C^a}^a\}$ the treatment-specific partition of the indices $\{1, \ldots, n^a\}$, where C^a is the number of clusters among patients treated with therapy a and $n_j^a = \mid S_j^a \mid$ is the cardinality of cluster j, for $j = 1, \ldots, C^a$. Finally, cluster-specific quantities are denoted with the super script "$\star$". For example, when considering the $j-$th cluster for treatment a, the response vector is $\boldsymbol{y}_j^{a\star} = \{y_i^a : i \in S_j^a\}$ while $\boldsymbol{x}_j^{a\star} = \{\boldsymbol{x}_i^a : i \in S_j^a\}$ is the partitioned covariate matrix. Using a conjugate prior for $\boldsymbol{\pi}_i^a$, we assume the following hierarchical model for $a = 1, \ldots, T$:

$$y_i^a | \boldsymbol{\pi}_i^a \stackrel{\text{ind}}{\sim} \text{Multinomial}(1, \boldsymbol{\pi}_i^a)$$

$$\boldsymbol{\pi}_1^a, \ldots, \boldsymbol{\pi}_{n_a}^a \mid \boldsymbol{\eta}_1^{a\star}, \ldots, \boldsymbol{\eta}_{C_a}^{a\star}, \rho^a, \boldsymbol{\beta} \sim \prod_{j=1}^{C^a} \prod_{i \in S^a} \text{Dirichlet}(\boldsymbol{\pi}_i^a; \boldsymbol{\gamma}_i^a(\boldsymbol{\eta}_j^{a\star}, \boldsymbol{\beta}, z_i^a)),$$

where $\boldsymbol{\beta} = (\boldsymbol{\beta}_1, \ldots, \boldsymbol{\beta}_K)$ is a $P \times K$ matrix of regression parameter shared across levels of response and individuals. The K-dimensional vectors $\boldsymbol{\eta}_1^{a\star}, \ldots, \boldsymbol{\eta}_{C_a}^{a\star}$ are cluster-specific parameters, that is, $\boldsymbol{\eta}_j^{a\star}$ is a parameter shared by all the individual in cluster S_j^a. Finally, $\boldsymbol{\gamma}_i^a(\boldsymbol{\eta}_j^{a\star}, \boldsymbol{\beta}, z_i^a) = (\gamma_{i1}^a(\eta_{j1}^{a\star}, \boldsymbol{\beta}_1, z_i^a), \ldots, \gamma_{iK}^a(\eta_{jK}^{a\star}, \boldsymbol{\beta}_K, z_i^a))$, is a vector of log-linear functions on the prognostic marker and cluster-specific parameters defined as follows:

$$\log(\gamma_{ik}^a(\eta_{jk}^{a\star}, \boldsymbol{\beta}_k, z_i^a)) = \eta_{jk}^{a\star} + \beta_{1k} z_{i1}^a + \cdots + \beta_{Pk} z_{iP}^a.$$

2.1 Priors

The choice of a covariate-dependent prior on the random partition enables predictive biomarkers to drive the clustering. Priors for $\{\rho^a\}$ and $\{\eta_j^{a\star}\}$, are defined independent across treatments. In fact, we want to allow the response probabilities to change from treatment to treatment even for subject with similar genetic profile. This independence assumption prevents the model from inducing a partition that implies the same response probability for genetically similar subjects that have received different treatments. The joint law of $(\rho^a, \eta_j^{a\star})$ is assigned hierarchically as:

$$P(\rho^a = \{S_1^a, \ldots, S_{C^a}^a\} \mid \boldsymbol{x}^a) \propto \prod_{j=1}^{C^a} c(|S_j^a|) g(\boldsymbol{x}_j^{a\star}), \tag{1}$$

$$\eta_1^{a\star}, \ldots, \eta_{C^a}^{a\star} \mid C^a \overset{\text{iid}}{\sim} p_0.$$

In Equation (1) the prior on the random partition is given via *cohesion* function c and *similarity* function g.

The cohesion function acts on clusters, depending only on the cluster size. Following [5], we choose a commonly adopted cohesion function, that is $c(S_j^a) = \alpha\Gamma(|S_j^a|)$, $\alpha > 0$, corresponding to the marginal partition model available from a Dirichlet process.

The *similarity* g is a non-negative function that measures how homogeneous patients in the same cluster are, with respect to predictive markers. It plays a crucial role since it increases the probability that patients with close genetic profiles are co-clustered. In Sect. 2.2 we list and describe two *similarity* function g along with strategies designed to temper the covariates' influence on clustering.

Following [10], for p_0 we adopted a conjugate Normal-Inverse Wishart. The posterior distribution for $\eta^{a\star}$ results in C^a independent multivariate normal densities.

The priors for the parameters $\{\boldsymbol{\beta}_k\}$ are assumed to be independent and, to enhance predictive performance, we specified horseshoe priors: $\beta_{pk} \sim N(0, \sigma_{pk}^2)$, for $p = 1, \ldots, P$, where $\sigma_{pk}^2 = \lambda_{pk}^2 \cdot \tau_k^2$, with $\lambda_{pk}, \tau_k \sim \text{HalfCauchy}(0, 1)$.

2.2 Similarity Function

Predictive biomarkers drive the clustering process trough the *similarity* function, that measure the homogeneity of the $x_i \in \boldsymbol{x}_j^\star$. In theory any non-negative function that produces larger values for more close covariates is suitable. In order to evaluate the influence of this choice on the response to treatment prediction we present the two *similarity functions* that are compared in the simulation study. As mentioned before, in order to counteract the strong effect that a large number of covariates may have on partition probabilities, we adopt a strategy to temper their effects. In particular, we briefly discuss the coarsening of the *similarity* function.

The original *similarity* function proposed by [5] is to choose $g(\boldsymbol{x}_j^{a\star})$ as the marginal probability of an auxiliary probability model. It takes the form

$$g(\boldsymbol{x}_j^{a\star}) = \int \prod_{i \in S_j^a} q(x_i^a \mid \boldsymbol{\xi}_j^{a\star}) q(\boldsymbol{\xi}_j^{a\star}) \mathrm{d}\boldsymbol{\xi}_j^{a\star}. \tag{2}$$

Note that $\{x_i^a\}$ are not considered random: this structure is convenient because the correlation induced by the cluster-specific parameters $\{\boldsymbol{\xi}_j^{a\star}\}$ leads to large values of $g(\boldsymbol{x}_j^{a\star})$ for close $\{x_i^a\}$.

For continuous covariates [5] suggests as default choice for $g(\boldsymbol{x}_j^{a\star})$ the marginal distribution of $x_j^{a\star}$ under a normal sampling model. A conjugate pair for $q(\cdot \mid \boldsymbol{\xi}^{a\star})$ and $q(\boldsymbol{\xi}^{a\star})$ greatly facilitates the evaluation of $g(\boldsymbol{x}_j^{a\star})$: $q(\cdot \mid \boldsymbol{\xi}_j^{a\star}) = N(\cdot | m_j^{a\star}, \upsilon_j^{a\star})$ and $q(\boldsymbol{\xi}_j^{a\star}) = q(m_j^{a\star}, \upsilon_j^{a\star}) = NIG(m_j^{a\star}, \upsilon_j^{a\star} | m_0, k_0, \upsilon_0, n_0)$, that are the Normal and Normal-Inverse-Gamma density functions, respectively. A simplified version of this conjugate model forces covariate clusters to have the same variance: $\upsilon_j^{a\star} = \upsilon^{a\star}$ and results in $q(\boldsymbol{\xi}_j^{a\star}) = N(m_j^{a\star} \mid m_0, s_0^2)$. We will refer to this latter formulation as the "Auxiliary NN" and the first one as the "Auxiliary NNIG". Note that we focus here on continuous covariates. A major advantage offered by similarities of the form of (2) is that they easily account also for categorical, ordinal and count covariates [6].

[9] propose a variation of (2), defining $g(\boldsymbol{x}_j^\star)$ as the posterior predictive distribution of $\boldsymbol{x}_j^\star$ in cluster S_j:

$$g(\boldsymbol{x}_j^{a\star}) = \int \prod_{i \in S_j^a} q(x_i^a \mid \boldsymbol{\xi}_j^{a\star}) q(\boldsymbol{\xi}_j^{a\star} | \boldsymbol{x}_j^{a\star}) \mathrm{d}\boldsymbol{\xi}_j^{a\star}, \tag{3}$$

with $q(\boldsymbol{\xi}_j^\star | \boldsymbol{x}_j^\star) \propto \prod_{i \in S_j} q(x_i | \boldsymbol{\xi}_j^\star) q(\boldsymbol{\xi}_j^\star)$. Since the covariates are used twice, this function is called "Double-dipper". The rationale for this formulation, that has the same form as (2), is to give more weight to the local covariate structure. This is pursued by weighting $\boldsymbol{x}_j^\star$s "likelihood" with the "posterior distribution" of $\boldsymbol{\xi}_j^\star$ instead of its "prior".

As for the *auxiliary similarity*, when x is continuous we can have the "Double-dipper NN" or "Double-dipper NNIG". Finally note that, for multivariate $\boldsymbol{x}_i = (x_{i1}, \ldots, x_{iQ})$, as in our case of study, we use $g(\boldsymbol{x}_j^\star) = \prod_q^Q g(\boldsymbol{x}_{jq}^{a\star})$.

As an alternative to variable selection or to reducing the dimensionality of the covariate space through the use of sufficient statistics, [8] proposes to calibrate the influence of covariates on clustering. In particular we consider the *coarsened similarity function*:

$$\tilde{g}(\boldsymbol{x}_j^{a\star}) = g(\boldsymbol{x}_j^{a\star})^{1/Q}. \tag{4}$$

In order to shrink the degree of *coarsening* we want to induce on the partition probabilities, we also consider a small variation of (4) which will be referred to as *shrunk coarsened similarity*: $\tilde{g}(\boldsymbol{x}_j^{a\star}) = g(\boldsymbol{x}_j^{a\star})^{1/\sqrt{Q}}$.

2.3 Posterior Computation

A MCMC procedure is used to fit the PPMx model. The core part of the algorithm is the updating of the cluster labels. The computation associated with fitting Equation (1) is based on [7]'s Algorithm 8, where applying a Gibbs sampling to a state augmented by the addition of auxiliary parameters greatly facilitates the update of the partition. Conditional on the updated cluster labels, all the remaining parameters are easily updated with Gibbs sampler or Metropolis-Hastings steps.

3 Treatment Selection

In order to select the optimal treatment for a new, untreated patient $\tilde{i}$, we are interested in the predictive probability of $y_{\tilde{i}}$. Given the observed responses for the n^a patients previously treated with therapy a, that is $\boldsymbol{y}^a$, the predictive probability of response level k under treatment a is

$$p(y_{\tilde{i}} = k \mid \boldsymbol{y}^a, \boldsymbol{z}^a, \boldsymbol{x}^a, \boldsymbol{z}_{\tilde{i}}, \boldsymbol{x}_{\tilde{i}}),$$

where $\boldsymbol{z}_{\tilde{i}}$ and $\boldsymbol{x}_{\tilde{i}}$ denote the P and Q dimensional vectors containing prognostic and predictive markers for the new patient. To facilitate treatment selection for multinomial ordinal outcomes, we adopt utility weights. In clinical oncology response categories are ordinal and consider changes in tumor size and/or distant migration after the treatment. We establish utility weights that turn a multinomial setting into a one-dimensional selection criterion considering the relative importance of each level of the ordinal response. Let $\boldsymbol{\omega}$ be a $K-$dimensional vector denoting the utility assigned to tumor response levels. To make $\boldsymbol{\omega}$ reflect clinical importance of each level (non respondent, partially respondent and respondent), we set $\boldsymbol{\omega} = (0, 40, 100)^\top$, following [4]. We can then compute the mean predictive utility for patient $\tilde{i}$ as:

$$\varphi^a(\tilde{i}) = \sum_{k=1}^{K} \omega_k p(y_{\tilde{i}} = k \mid \boldsymbol{y}^a, \boldsymbol{z}^a, \boldsymbol{x}^a, \boldsymbol{z}_{\tilde{i}}, \boldsymbol{x}_{\tilde{i}}).$$

The $\tilde{i}-$th patient will be assigned to the therapy ensuring the largest predictive utility, that can be considered to be optimal among the competing treatments.

4 Illustrative Example

To empirically assess the performance of the coarsened similarity function presented in Sect. 2.2, we conduct a simulation study. To compare model fit and treatment selection we generate synthetic data adopting the processes designed by [4] (see Scenario 2), with the only difference that we use 10 predictive markers (instead of 90), while we consider the same two prognostic covariates.

This procedure yields $n = 152$ patients that are assigned to $T = 2$ competing treatment. We consider 3 levels for the ordinal-valued response variable. We standardize all predictive biomarkers.

The hyperparameters for Auxiliary NN and Double-dipper NN similarities are $(m_0 = 0, s_0^2 = 1)$. For hyperparameters needed when the NNIG model is employed in the similarities, on the ground of the results obtained by [8] in their extensive simulation study and sensitivity analysis, we set $(m_0 = 0, k_0 = 1, v_0 = 10, n_0 = 2)$.

For each similarity function we run the PPMx for 150,000 iterations, descarding the first 50,000 due to burn-in and keeping each 10−th draw from the posterior distribution. To compare the goodness-of-fit we report the log pseudomarginal likelihood (LPLM). To evaluate the predictive performances we adopt the same metrics as in [4]:

(i) MOT, that is the number of misassigned patients;
(ii) $\%\Delta$MTU, it measures the relative gain in treatment utility with respect to the other treatment; note that it is defined only for the case of two alternative treatments. It ranges from -1 to 1 ($\%\Delta$MTU $= 1$ only in the case of optimal treatment assignment rule);
(iii) NPC that is the number of correctly predicted outcomes.

Prediction is based on a leave-one-out cross-validated strategy. The numerical results reported in Table 1 are averaged over 100 data sets generated for each case. Standard deviations are given in brackets. The best performance for each metric is reported in bold.

The Double-dipper similarity outperforms the Auxiliary similarity function. Double-dipper best performances are probably due to the larger weight given to the covariates in the model-based clustering process.

Focusing on the lower pane of Table 1, we notice that the Double-dipper function delivers better results when the NNIG model is assumed. In fact, NNIG offers a greater flexibility than NN, as it does not force clusters to share the same variance.

Restricting our focus to the Double-dipper NNIG similarity function, Table 1 offers a last comparison between Coarsening and Shrunk Coarsening. The former achieves better performances in terms of goodness-of-fit, while the latter is to be preferred according to those metrics evaluating prediction. Shrunk Coarsened similarity outperforms Coarsened similarity assigning fewer patient to the non optimal treatment (15.18 vs 24.13) and reaching a larger relative gain in treatment utility (82% versus 64%). Coarsening, on the other hand, yields slightly better performances in terms of number of correctly predicted outcome and LPML.

Given the focus on treatment selection rather than inference on model parameters, Shrunk Coarsened Double-dipper NNIG is the similarity function best suited for our model.

Table 1 Simulation study on *similarity functions*

Similarity	MOT	%ΔMTU	NPC	LPML
Coarsened Auxiliary NN	34.33	0.47	80.63	−129.98
	(4.71)	(0.05)	(6.06)	(4.33)
Coarsened Auxiliary NNIG	28.50	0.58	80.41	−129.18
	(5.79)	(0.08)	(5.92)	(4.47)
Shrunk Coarsened Auxiliary NN	55.70	0.30	74.28	−155.51
	(33.72)	(0.41)	(7.00)	(3.81)
Shrunk Coarsened Auxiliary NNIG	70.82	0.10	67.00	−156.75
	(7.64)	(0.10)	(6.75)	(3.95)
Coarsened Double-dipper NN	31.93	0.50	79.91	−124.08
	(4.71)	(0.05)	(5.95)	(4.17)
Coarsened Double-dipper NNIG	24.13	0.64	**81.70**	**−121.26**
	(6.66)	(0.09)	(5.87)	(3.65)
Shrunk Coarsened Double-dipper NN	19.83	0.73	77.40	−141.58
	(9.03)	(0.10)	(6.50)	(4.42)
Shrunk Coarsened Double-dipper NNIG	**15.98**	**0.82**	77.08	−146.17
	(8.06)	(0.09)	(6.05)	(4.44)

5 Conclusion

Employing PPMx to cluster together patients with close genetic profiles and then evaluate the effectiveness of competing treatments on groups of similar patients shows promise. In this paper we focus on the choice of the similarity function, that is pivotal in PPMx models, in the framework of optimal treatment selection. We find the Double-dipper similarity to perform particularly well when a shrunk coarsening is employed and the NNIG model is adopted.

Several extension are currently under investigation, with a sharp focus on similarity functions that could enable us to include a larger number of predictive markers.

References

1. Golub, T.R., Slonim, D.K., Tamayo, P., Huard, C., Gaasenbeek, M., Mesirov, J.P., Coller, H., Loh, M.L., Downing, J.R., Caligiuri, M.A.: Molecular classification of cancer: class discovery and class prediction by gene expression monitoring. Science **286**, 531–537 (1999)
2. Kosorok, M.R., Laber, E.B.: Precision medicine. Annu. Rev. Stat. Appl. **6**, 263–286 (2019)
3. Ma, J., Stingo, F.C., Hobbs, B.P.: Bayesian predictive modeling for genomic based personalized treatment selection. Biometrics **72**, 575–583 (2016)
4. Ma, J., Stingo, F.C., Hobbs, B.P.: Bayesian personalized treatment selection strategies that integrate predictive with prognostic determinants. Biometrical J. **61**, 902–917 (2019)
5. Müller, P., Quintana, F., Rosner, G.L.: A product partition model with regression on covariates. J. Comput. Graph. Stat. **20**, 260–278 (2011)

6. Müller, P., Quintana, F., A., Jara, A., Hanson, T.: Bayesian Nonparametric Data Analysis. Springer, Heidelberg (2015)
7. Neal, R.M.: Markov chain sampling methods for Dirichlet process mixture models. J. Comput. Graph. Stat. **9**, 249–265 (2000)
8. Page, G.L., Quintana, F.: Calibrating covariate informed product partition models. Stat. Comput. **28**, 1009–1031 (2018)
9. Quintana, F., Müller, P., Papoila, A.L.: Cluster-specific variable selection for product partition models. Scand. J. Stat. **42**, 1065–1077 (2015)
10. West, M., Müller, P., Escobar, M.D.: Hierarchical priors and mixture models, with applications in regression and density estimation. In: Aspects of Uncertainty: A Tribute to D.V. Lindley, pp. 363–386 (1994)

Bayesian Growth Curve Model for Studying the Intra-abdominal Volume During Pneumoperitoneum for Laparoscopic Surgery

Gabriel Calvo, Carmen Armero, Virgilio Gómez-Rubio, and Guido Mazzinari

Abstract Laparoscopy is a surgical procedure carried out in the abdomen or pelvis through small incisions with the help of a camera to view the organs in the abdomen or permit small-scale surgery. This technique needs the abdomen to be insufflated with carbon dioxide (CO_2) to obtain a working space for surgical instruments' manipulation. Identifying the critical point at which insufflation should be limited is crucial to maximizing surgical working space and minimizing injurious effects. Bayesian nonlinear growth mixed-effects models are applied to data coming from a repeated measures design. The study allows to assess the relationship between the insufflation pressure and the intra-abdominal volume as well as to draw inferences and predictions for the main outcomes of the process.

Keywords Intra-abdominal pressure · Logistic function · Sigmoidal function

1 Introduction

Laparoscopy is an operation carried out in the abdomen or pelvis through small incisions with the help of a camera. It is performed by insufflating CO_2 into the abdomen that yields a working space, i.e., pneumoperitoneum, and passing surgical instruments through small incisions using a camera to have external visual control

G. Calvo (✉) · C. Armero
Universitat de València, Carrer Doctor Moliner 50, 46100 Burjassot, Spain
e-mail: Gabriel.Calvo@uv.es

C. Armero
e-mail: Carmen.Armero@uv.es

V. Gómez-Rubio
Universidad de Castilla-La Mancha, Avda. de España s/n, 02071 Albacete, Spain
e-mail: Virgilio.Gomez@uclm.es

G. Mazzinari
Hospital Universitari i Politècnic la Fe, Avda. de Fernando Abril Martorell 106, 46026 Valencia, Spain

R. Argiento et al. (eds.), *New Frontiers in Bayesian Statistics*, Springer Proceedings in Mathematics & Statistics 405, https://doi.org/10.1007/978-3-031-16427-9_11

of the procedure [7]. Laparoscopy technological development has been limited to improvements in camera image quality, whereas little innovation has been made in insufflation devices.

This paper is based on a previously published work [1] about the subject with the aim of estimating, through Bayesian inference, a non-linear model [3] about the relationship between the CO_2 insufflation pressure, i.e., intra-abdominal pressure (IAP), measured in mmHg, and the intra-abdominal volume (IAV) generated, measured in L.

Data for the current modelling come from a previously published individual patient meta-analysis [6] that included experimental information from three previous clinical studies. It consists in a repeated measure design where the variable of interest IAV is measured for each individual with regard to different IAP values, and age and sex have been taken into account as covariates. The final databank has 198 patients, 118 men and 80 women, with a total of 1361 observations.

2 Bayesian Growth Curve Model

Let the non-linear mixed effect model for the random variable Y_{ij} that records the IAV value for individual $i, i = 1, \dots, n$ with IAP value $x_{ij}, j = 1, \dots, J_i$, defined as

$$(Y_{ij} \mid \mu_{ij}, \sigma^2) \sim \mathrm{N}(\mu_{ij}, \sigma^2), \tag{1}$$

where σ^2 is the unknown variance associated to the random measurement error. The mean μ_{ij} is the true IAV value of patient i with IAP value x_{ij}, and can be expressed in terms of the conditional logistic growth function [1] as follows

$$(\mu_{ij} \mid a_i, b_i, c_i, x_{ij}) = \frac{a_i}{1 + \exp\{-(b_i + c_i\, x_{ij})\}}, \tag{2}$$

where

$$a_i = \beta_0^{(a)} + u_i^{(a)} + \beta_W^{(a)} I_W(i) + \beta_A^{(a)} Age_i, \tag{3}$$

$$b_i = \beta_0^{(b)} + u_i^{(b)} + \beta_W^{(b)} I_W(i) + \beta_A^{(b)} Age_i, \tag{4}$$

$$c_i = \beta_0^{(c)} + \beta_W^{(c)} I_W(i). \tag{5}$$

$\boldsymbol{\beta}_0 = (\beta_0^{(a)}, \beta_0^{(b)}, \beta_0^{(c)})'$ stands for the common intercept with the men patients being the reference group, $I_W(i)$ is the indicator variable with value 1 if individual i is a woman and 0 otherwise, with associated coefficients $\boldsymbol{\beta}_W = (\beta_W^{(a)}, \beta_W^{(b)}, \beta_W^{(c)})'$, and $\boldsymbol{\beta}_A = (\beta_A^{(a)}, \beta_A^{(b)})'$ are the vector of regression coefficients associated with the age.

Random effects $u_i^{(a)}$ and $u_i^{(b)}$, $i = 1, \ldots, n$, are assumed normally distributed according to $f(u_i^{(a)}|\sigma_a^2) = \mathrm{N}(0, \sigma_a^2)$ and $f(u_i^{(b)}|\sigma_b^2) = \mathrm{N}(0, \sigma_b^2)$, respectively.

In addition, the Bayesian model is completed with the elicitation of a prior distribution for the parameters and hyperparameters $\boldsymbol{\theta} = (\boldsymbol{\beta}_0, \boldsymbol{\beta}_W, \boldsymbol{\beta}_A, \sigma, \sigma_a, \sigma_b)'$ of the model. We assume prior independence and a non-informative prior scenario for all of them. Normal distributions with a large standard deviation for most of the common regression coefficients are selected as prior distributions, $\pi(\beta_0^{(a)}) = \pi(\beta_0^{(b)}) = \pi(\beta_0^{(c)}) = \pi(\beta_W^{(a)}) = \pi(\beta_W^{(b)}) = \pi(\beta_W^{(c)}) = \pi(\beta_A^{(a)}) = \pi(\beta_A^{(b)}) = \mathrm{N}(0, 10^2)$. Furthermore, following [4], we have selected wide uniform distributions as prior distributions for all standard deviation parameters, $\pi(\sigma) = \pi(\sigma_a) = \pi(\sigma_b) = \mathrm{U}(0, 10)$, as well as for $\beta_0^{(a)}$ and $\beta_0^{(c)}$, $\pi(\beta_0^{(a)}) = \mathrm{U}(0, 20)$, and $\pi(\beta_0^{(c)}) = \mathrm{U}(0, 10)$.

An important value for clinical practice is the asymptotic deceleration point (ADP) [5]. This point is the last of the critical points of the logistic growth curve. It is calculated by equalling the fourth derivative to 0. It is located after the maximum deceleration point, which is the minimum of the second derivative of the curve. The ordinate $(a_i(3 + \sqrt{6})/6)$ of point ADP for individual i is extremely close to the horizontal asymptote of the curve a_i, so its second derivative, i.e. its acceleration, is negative but already very close to 0. This is due to the fact that the logistic growth curve is always increasing. Finally, the value of the abscissa at this point is expressed as

$$ADP_x(i) = -(\ln(5 - 2\sqrt{6}) + b_i)/c_i. \qquad (6)$$

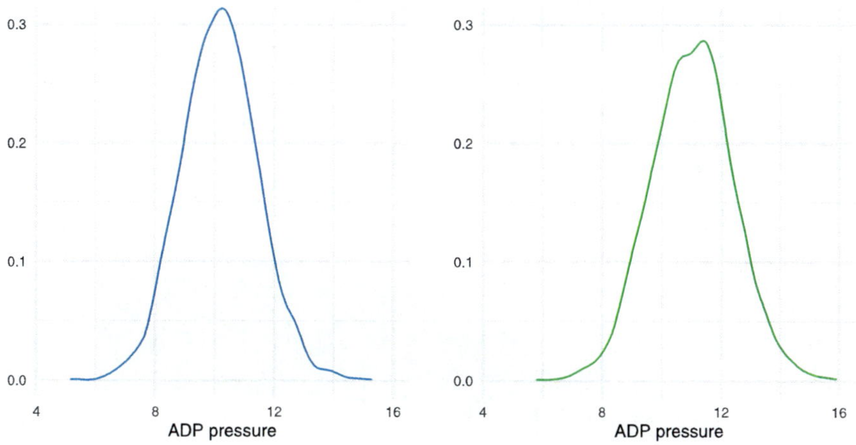

Fig. 1 Approximate posterior distribution density of the ADP pressure for men (on the left) and women (on the right)

3 Posterior Results

The posterior distribution $\pi(\boldsymbol{\theta}, \boldsymbol{u} \mid \mathcal{D})$, where $\boldsymbol{u} = (u_i^{(a)}, u_i^{(b)})$ and $\mathcal{D}$ represents the observed data, contains all the relevant information of the problem and it is usually the starting point of all relevant inferences. It was approximated by means of Markov chain Monte Carlo (MCMC) simulation methods through the JAGS software [8].

Figure 1 shows the approximate posterior distribution density of the ADP_x for men and women aged 64.65 years (the mean of the sample). Posterior mean for the ADP_x's is 8.86 mmHg for men and 10.06 mmHg in the group of women. Men do not need as much pressure as women to obtain the optimal IAV. This relevant difference between male and female patients should be taken into account during the laparoscopic procedure.

Individual prediction is a relevant issue of the study. In this sense, the posterior predictive distribution of the IAV variable $Y_{n+1,j}$ of a new individual of the target population with an IAP value $x_{n+1,j}$ is computed as follows

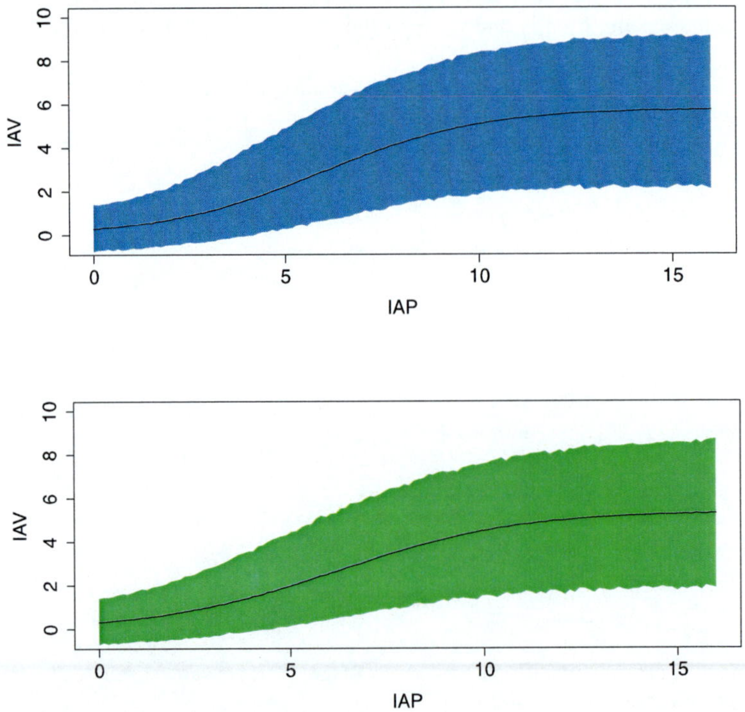

Fig. 2 Posterior predictive mean of the IAV and 95% predictive interval with regard to IAP values for a man (top panel) and a woman (bottom panel) aged 64.65 years (the sample mean)

$$(Y_{n+1,j} \mid x_{n+1,j}, \mathcal{D}) \sim \int (Y_{n+1,j} \mid x_{n+1,j}, \boldsymbol{\theta}, \boldsymbol{u}_{n+1})\, \pi(\boldsymbol{\theta}, \boldsymbol{u}_{n+1} \mid \mathcal{D})\, \mathrm{d}(\boldsymbol{\theta}, \boldsymbol{u}_{n+1}), \tag{7}$$

where $\mathcal{D}$ represents the observed IAV data.

Figure 2 shows the posterior predictive mean and a 95% predictive interval for the IAV value of a new individual of the target population with respect to their IAP by sex. The stabilisation of the values of IAV in both groups can be clearly seen, as well as the variability associated with the predictive processes.

4 Conclusions

We have presented a logistic growth model that aims to achieve an optimal surgical workspace while minimizing the pressure administered to the patient. According to our results, the pressure needed to arrive to the ADP, which is the most critical point in this work, is higher for women. However, there is high variability in the posterior results.

This is a preliminary Bayesian study that will serve as a starting point for testing more complex models to better explain the data. Additionally, in the future, other covariates related to anthropometric measurements will be recorded and included into the analysis to reduce the uncertainty about the estimates and predictions, and increase the accuracy of the insufflation procedure. In addition, to better explain heterogeneity among individuals, an alternative treatment of random effects in the non-parametric statistical scheme [2] would be an interesting project.

Acknowledgements This paper was supported by research grant PID2019-106341GB-I00 funded by Ministerio de Ciencia e Innovación (Spain) and the Project MECESBAYES (SBPLY/17/180501/000491) funded by the Consejería de Educación, Cultura y Deportes, Junta de Comunidades de Castilla-La Mancha (Spain). Gabriel Calvo is also supported by grant FPU18/03101 from the Ministerio de Ciencia e Innovación (MCI, Spain). Merck Sharp & Dohme funded the IPPColLapse II study (Protocol Code No. 53607).

References

1. Calvo, G., Armero, C., Gómez-Rubio, V., Mazzinari, G.: Bayesian hierarchical nonlinear modelling of intra-abdominal volume during pneumoperitoneum for laparoscopic surgery. Stat. Op. Res. Trans. **45**(2), 143–162 (2021)
2. Cruz-Mesía, R.D.L., Quintana, F.A., Müller, P.: Semiparametric Bayesian classification with longitudinal markers. J. Roy. Stat. Soc. **56**(2), 119–137 (2007)
3. Davidian, M.: Non-linear mixed-effects model. In: Fitzmaurice, G., Davidian, M., Verbeke, G., Molenberghs, G. (eds.) Longitudinal Data Analysis, pp. 121–156. CRC Press (2008)
4. Gelman, A.: Prior distributions for variance parameters in hierarchical models (comment on article by Browne and Draper). Bay. Anal. **1**(3), 515–534 (2006)

5. Korkmaz, M., Volkan, O.D.A., Basustaoglu, E.O.: A study over determination of asymptotic deceleration and absolute acceleration points in logistic growth model. Turk. J. Philos. Math. Comput. Sci. **10**, 33–37 (2018)
6. Mazzinari, G., Diaz-Cambronero, O., Serpa Neto, A., Martínez Cañada, A., Rovira, L., et al.: Modeling intra-abdominal volume and respiratory driving pressure during pneumoperitoneum insufflation—a patient-level data meta-analysis. J. Appl. Phys. **130**(3), 721–728 (2021)
7. Neugebauer, E.A.M., Becker, M., Buess, G.F., et al.: EAES recommendations on methodology of innovation management in endoscopic surgery. Surg. End. **24**(7), 1594–1615 (2010)
8. Plummer, M.: JAGS: a program for analysis of Bayesian graphical models using Gibbs sampling. In: Plummer, M., Hornik, K., Leisch, F., Zeileis, A. (eds.) Proceedings of the 3rd International Workshop on Distributed Statistical Computing, pp. 1–10 (2003)

Author Index

R. Argiento et al. (eds.), *New Frontiers in Bayesian Statistics*, Springer Proceedings in Mathematics & Statistics 405, https://doi.org/10.1007/978-3-031-16427-9